An Invitation to Cognitive Science

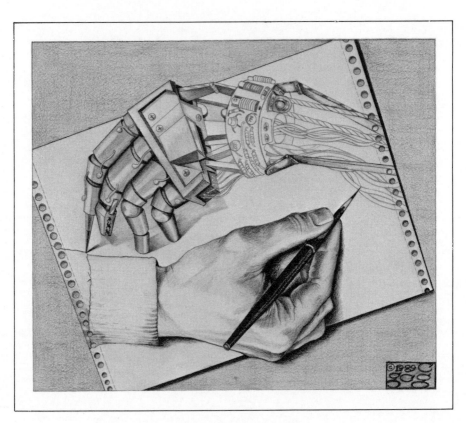

Drawing by Christine Ciesiel after M.C. Escher.

An Invitation to Cognitive Science

Justin Leiber

Basil Blackwell

Basil Blackwell, Inc.
3 Cambridge Center
Cambridge, Massachusetts 02142, USA

Basil Blackwell Ltd
108 Cowley Road, Oxford, OX4 1JF, UK

Lines from 'The Circus Animals' Desertion' by W. B. Yeats are reprinted with permission of Macmillan Publishing Company from *The Poems of W. B. Yeats: A New Edition*, edited by Richard J. Finneran. Copyright 1940 by Georgie Yeats, renewed 1968 by Bertha Georgie Yeats, Michael Butler Yeats, and Anne Yeats.

Line from 'Among School Children' by W. B. Yeats is reprinted with permission of Macmillan Publishing Company from *The Poems of W. B. Yeats: A New Edition*, edited by Richard J. Finneran. Copyright 1933 by Macmillan Publishing Company, renewed 1961 by Bertha Georgie Yeats.

Library of Congress Cataloging in Publication Data
Leiber, Justin.
An invitation to cognitive science/Justin Leiber.
p. cm.
Includes bibliographical references and index.
ISBN 0-631-17004-9 —ISBN 0-631-17005-7 (pbk.)
1. Cognitive science. I. Title.
BF311.L423 1991
153—dc20   90-1240
72267   CIP

British Library Cataloguing in Publication Data
A CIP catalogue record for this book is available from the British Library.

Typeset in 10 on 12pt Imprint
by Photo·graphics, Honiton, Devon
Printed in Great Britain

Contents

Preface

*Like the choice between competing political institutions, that between com-
peting paradigms proves to be a choice between incompatible modes of
community life. Because it has that character, the choice is not and cannot
be determined merely by the evaluative procedures of normal science, for
these depend in part upon a paradigm and that is at issue.*

Thomas Kuhn[1]

This book paints an idiosyncratic and personal picture. So do most books,
that is really what they are for, though nearly all get comfortably into
their stride as the reader allows himself to forget the narrator, so that
the story, the arguments, the data, the records, the instructions, the
programs, the artefacts, now appear to speak for themselves. Literature
calls this readerly allowance 'willing suspension of disbelief' and so, as
literary fiction, it strives for an inner coherence that vividly and instructively shows what human possibilities can comprise, given some stipulated
setting. At farthest remove from this we find the census taker or obituary
writer, who aims only to have correct particular names, domiciles, birth
dates, death dates, occupations, and family and marital status. Sciences,
technologies, professions, and other arts trail every which way between
these narrative extremities and several others.

The more specific, untheoretical and established the field of knowledge
or technique, the more irrelevant is its *actual* historical construction by
persons, proposals, and foundational texts and inventions: practitioners
can write ahistorical anonymous handbooks and standard, noncontentious
introductory texts. In cognitive science, however, we have several separate
disciplines jostling each other for position in a coalescence that may be
illusory. Worse yet, each of these disciplines is more well established and
has a stronger sense of its own voice and commitments than anything
cognitive science as a larger discipline puts forward with assurance. A
serious introductory textbook of cognitive science cannot avoid being a
contentious manifesto. Even more dubious would be a cacophony in
which each discipline has its chapter, for such text divisively becomes
just that: a *mélange* of introductions to different disciplines.

Even an *invitation* is suspect.

My proposal is to let the proper authors of this invitation speak through my commentary, which is, as narratives explicitly and other texts implicitly, intended to simulate a kind of consciousness, a way of looking at matters, a point of view. The characters of my narrative are the authoritative proposals, inventions, proofs, and texts that voice this invitation, sprawling across disciplines: as any novelist knows it is more convincing to let the characters speak for themselves and, as any investigator, historian, or lawyer knows, original texts are the best point of departure. What I mean by character, here, is absorbed into the specific proposal or formula or text or invention in its relationship to others within an apparently developing and coalescing enterprise, though a merger still shaken within and pressured by our surrounding culture.

In whatever way cognitive science develops, these authorial inventions will be part of the story, however differently they will be read some decade hence. Your invitation is first hand.

If I give a narrative account of cognitive science in a look through its classical formulations, I justify my title. For indeed I invite you to these texts and inventions, both to exhibit cognitive science through them, and to persuade you to further wonder and worry about these creations and the prospect they open for us, the understanding of ourselves.

Though this account is idiosyncratic and personal, a textual, and individual, view, this is not for want of help and counsel. A. J. Ayer, Steve Branson, Noam Chomsky, Daniel Dennett, Barbara Foorman, Nick Fotion, Jim Garson, Jay Hullett, Kathryn Hume, Craig Lawson, Stanley Martens, Marvin Minsky, Hilary Putnam, and others have helped me with suggestions and corrections.

Though its roots are as old as systematic inquiry, our subject seems so new that you might well think cognitive science erupted into the world complete, a few years back, like Athena springing fully adult from Zeus' head (or so Dorian Greek barbarians legitimized their conquest of a more matriarchal civilization). Even the field's practitioners can fall into this view. For example, participants in the explosive current debate between 'connectionist' models of human cognition and serial, programmatic models are unfamiliar with Turing's commanding 1950 exposition of their supposedly spanking new dispute. Similarly, though 'Turing Machine', 'Universal Turing Machine', and the 'Turing test' belong to the basic vocabulary of current cognitive science, practitioners learn them second hand: as we shall see, the original formulations can differ, strikingly and revealingly. Looking even farther back, it helps to understand that much of cognitive science can be understood as an attempt to make empirical and technical sense of the centuries-old clash between empiricist and rationalist models of cognition.

As perhaps the first cognitive scientist, Alan Turing's commanding formulation of the field's concepts, problems, and prospects invite us to make his prospectus our route for understanding. I have juxtaposed Turing with Ludwig Wittgenstein, whose writings give us a commanding, and devastating, critique of the autonomous, conscious mind through spell-binding portraits of the contradictions latent in our everyday cognitive life, the anomalies which have proved to be the paradigmatic routes of advance for the new cognitive science. The works of these two have provided the way of writing this book, a way intended to display such texts in terms of their paradoxical relationship to each other, and as they represent the central problems and achievements of cognitive science. If Turing sets the stage for the computational mind, Noam Chomsky's linguistics perhaps most tellingly represent the direct study of human cognition as a formal system and virtual machine – natural language as key to the formal properties of thought.

Athena's birth also expresses a deeper, more pervasive and systematic, blindness. We assume pridefully that human cognition is the only real natural cognition, the standard, the very stuff, to which, in small ways, some animals may aspire and to which anything else, certainly anything we build, is only simulation or copy. But if we see matters this way we must blind ourselves to the chordate mammal-specific, species-specific aspects of our cognition. Worse still, when someone shows us an aspect or system of human cognition that looks to be species-specific, we are now encouraged not to see what he is talking about, or not see it as cognitive at all but as fleshy, reflexive, or decorative. And this denial of ourselves, like Athena's, robs us of a full purchase on the question of how variegated thinking things might be; equally, we cannot then expect anything like a clear grasp of ourselves, of our cognitive virtues and vices, limitations and graces, homely peculiarities and glandular prejudices.

Early in this century, Otto Neurath wrote that, as humans attempting to understand our world and ourselves *in the structures of our natural language*, we cannot start from the beginning, building our ship in dockyard rather than having to manage the task of reconstruction and repair while at sea.[2] The cognitive scientist Alan Turing asks for licence to construct auxiliary craft, both to learn how to build simpler models from scratch and to give us a way to stand off and see better the larger dimensions and baroque peculiarities of our ship. A linguist might add, to deepen and make more realistic Neurath's metaphor, that the myriad ships are the individual speakers, who fall into separate schools by

virtue of their superficial coloring and decorations, and yet who all share
variations on a complex common underlying architecture – a below-deck,
structural architecture, indeed, so firm that you can hold up to view,
occasionally for improvement though, necessarily, mostly for understanding, some one plank or another, and rate one as unavoidable nature,
another as a grówth option in a naturally limited range, and a third as
decorative choice, recognizing it in all as the robust product of biological
growth; while the superstructures, the provisional and quite various
constructions of art and science, may be endlessly remodelled though the
ship is at sea.

1
Shake Hands

Nature imitates Art.

Oscar Wilde

How can we know the dancer from the dance?

William Butler Yeats

The fundamental invitation of cognitive science, *know thyself!*, is ancient, ardent, engaging, and enduring, as commanding and challenging as when first issued. The Socratic invitation is also, to speak more contentiously and ambiguously, universal, awful, and paradoxical. And, through the dazzling inventions and discoveries of the past few decades that converge in making cognitive science a genuine and burgeoning enterprise, an invitation that is now no vague piety but the marching orders of a large, partially structured but harmonically anarchic and quarrelsome legion – a legion, unlike the military sort, that deals in consciousness and thought, rather than death and obedience.

The invitation is universal in three senses. It has always beckoned us, even though we today can hope for more detailed enlightenment than could Socrates. But it also asks for what is common to us all as reasoners, perceivers, knowers, as functioning mental beings, regardless of the peculiarities of our physical bodies. Universal, perhaps further, in the sense that the account of our cognition will illustratively contrast what in our cognition might be common to all possible thinkers and what is class- or species-specific, so marking us out as one kind among many thinkers. And the invitation is awful in the original sense, just the sort of thing to make the hairs on the back of your neck stand up.

The invitation is also paradoxical because your mind seems to be the one thing that you cannot fail to know: it is what you have to see when you open your mind's eye, when, more tersely, you are conscious, and what does not then appear, open to clear view, cannot be part of your mind. My mind is the one thing *I* cannot be *told* to learn more about because my mind is what *I am, what I have learned, what I know*. A part of our language repeats this homely conviction to us endlessly; and yet one is so often made painfully aware that one understands oneself least of all. Indeed, our attempt to make computers *think* and *think like us* presents perhaps really only one great difficulty – our near abysmal ignorance of what we do when we think, of what we are actually doing when we win chess games or successfully maneuver cars through speeding traffic, hear what someone says in our native language or hit a moving target with a rock, recognize the face of a 50-year-old woman in her first-grade class photograph or call to mind a name we had forgotten.

Plato ennobled and deepened the paradox through ascribing to Socrates a knowledge that distinguished him from other Athenians: 'I know that I know nothing'. Since our understanding of mind has mostly grown in progressive realizations of the structure of our ignorance and of the magnitude of what demands explanation, and so has only just begun to build on this groundplan, I first invite you to a course of paradoxes, keys to the structure of cognition, language, perception, and consciousness, the *seams* of the explosive outpouring of our mental faculties. Aside from picturing creations that force cognition or formal system to reveal its structure, our account will range through the larger paradoxes, the foundational articles and books, that animate cognitive science, that force the structure of our sciences, and so of our self-cognition, into new array.

Do you read me?

Epimenides, a legendary Greek poet from Crete, surely began something with the statement, 'All Cretans are liars', though St Paul made the paradox ironclad by the lawyerly addition of '*always*'. One of themselves, even a prophet of their own, said, 'the Cretans are *always* liars', St Paul completed his sentence, in all too human disdain for paradoxes and puns, 'evil beasts, slow bellies'. You may find yourself muttering these dark last words of St Paul's as you read this book.

St Paul cursed profoundly, for paradoxes show us what is *unspeakable*, *impossible*, *uncountable*, *unprovable*, *undecidable*, *unhaltable* and *unmakeable*. To put it Paulinely, as heaven loves a sinner saved, so absurdity is the road to truth. Epimenides began with the *unspeakable*.

If what Epimenides says is true, then it is false. But if it is false, then of course it is true. But if it is true, then . . . The same problem arises more tersely with, 'I am now lying' or 'This sentence is false'. One could go shuttling along in such 'strange loops' forever, or perhaps simply conclude that some sentences say something both true and also false at the same time. Either alternative is unsatisfactory.

As a practical matter, most of us quarantine off such sentences as paradoxes, as we do puns more generally. We call puns the '*lowest* form of humour' because they dirty and profane rational discourse, jerking us rudely over the inhibitions that allow us to think, speak, and listen coherently, without being caught up in irrelevant ambiguities. Paradoxes are the most potent of puns, not alternate readings of innocent sentences suddenly forced upon us in context, but 'evil beast' sentences that have no nondestructive reading.

Paradoxes are fraudulent, perhaps impossible, sentences, with no coherent formal interpretation. We do not use them in our everyday reasoning: we had best not, for they would spread, like some particularly malignant computer virus, eventually establishing that every English sentence is both true and false.

I still vividly remember feeling nauseous the first time I was caught by such a paradox, going round and round, being certain there must be a way out and not finding one. One imagines that someone with a peculiarly limited and paranoid cast of mind might very well 'strange loop' along until some kind of physical collapse ensued, perhaps just a little like the epileptic seizure that can be induced in most of us by a stroboscopic light whose firing rate is determined by feedback from our brain waves.[1] Fledgling calculators, computers, and computer programs have been, of course, much more prone than we to fall into such spasms. (The problem is not just one that only plagues a sentence that refers to itself. Consider two sentences, one named 'Para', the other, 'Dox'. Para says 'Sentence Dox is false' and Dox says 'Sentence Para is false'. You can chain three sentences into paradox this way, or as many more as you like.)

Indeed, related paradoxes threaten individuals and collections, or sets, of individuals. The ancient Greeks were probably even more impressed by the paradox of 'Nothing is!' But if nothing *is*, then it is something, not just nothing. But if it *is something* then it is not (just) nothing. So nothing is not (just) nothing; nothing is not. But that tells us just exactly what nothing must be, namely, what is not. So nothing *is*. And so some ancient Greeks, understandably, labeled nothing 'unspeakably impossible' and strongly recommended not even mentioning it.

In *Through the Looking-Glass*, Lewis Carroll gives a version of a paradox nearly as old as Epimenides. The Red King exists simply as something dreamed by Alice. But this dreamed Red King is asleep and he is dreaming Alice: she exists simply as something dreamed by the Red King.

The artist, E. M. Escher, gave a visual version of this last paradox. One hand, with a pencil, is drawing another; the paradox arises when we see that the drawn hand is itself a looking glass: it is drawing the first hand. The self-denying spiral ensues: the first hand – pick whichever one you wish – is drawing, is bringing into existence, the other, which is what it is and indeed exists *because* the first hand is drawing it. But the second hand occupies the same position: it draws, and so brings into existence, the first hand. Each hand has equal claim to the label: the drawer; and each to the contrary label: the drawn.

We might call these two 'strange loops', Alice and the Red King in Carroll's narration and the two hands of Escher's drawing, *creator/creation* loops. In Escher's drawing the two hands have equal claim to both positions: that is what makes the scene depicted impossible. A real human hand, directed by a human artist's intelligence, can draw (or sculpt or manufacture or understand) a hand: that is possible, and so we can realistically picture such a scene. But the drawn (sculpted, manufactured, understood) hand cannot in turn *also* be a real human hand, directed by another human intelligence, drawing (sculpting, manufacturing, understanding) the first hand. As we look at Escher's drawing, we shuttle back and forth from these two ways of filling the *creator/creation* slots; we *cannot* have it both ways, though that, of course, is what Escher's drawing demands.

Similarly, a real human being, Alice, can dream (create, describe, understand) a Red King, whom she dreams to be dreaming of a girl named Alice. *That* situation is perfectly possible and nonparadoxical. Equally, a real human Red King (perhaps his given name is Eric, but his grumpy penchant for violence has given him this nickname) might dream up (create, describe, understand) a charming young daughter of the dean of an Oxford college, Alice, who, so the Red King dreams, dreams of a Red King. This is equally possible and nonparadoxical. The difference comes in the waking-up part.

In the first case, when Alice wakes, the dreamed Red King disappears; but if Alice dreams that the Red King wakes, Alice does not disappear. In the second case, when the Red King wakes, the dreamed Alice disappears, except perhaps in some tyrant-troubling memory of innocence; while should the Red King dream that his dream Alice awakens from her dream of him, at most we can only imagine the King suffering some brief physiological spasm, the substrait of a nightmarish turn in his imagination.

But Lewis Carroll's paradoxical vision is that Alice and the Red King each exist by virtue of the other's dream. If Alice awakes, the Red King *and his dream of Alice* disappear, so Alice disappears. And contrariwise, as the author of Alice would say: if the Red King wakes, Alice *and her dream of the Red King* disappear, so the Red King disappears. A little frivolously, one could say that the moral of this story, for Alice and the Red King, is that they ought to agree not to wake up.

Now there is some genuine illumination, a vivid sense of a distinction that hews everything right, in the Alice/Red King story. What makes the story self-destructively paradoxical is the causal stance, the insistence that in reality the creator/creation relationship can go only one way: that, brutally, either Alice *or* the Red King is *the real physical/neurological organism* that dreams (sculpts, narrates, understands) the other. Most of us would say, in concert with recent experimental research, that our dreams approximate the stream of consciousness that constitutes our everyday experience of the world. We endorse Descartes' view that while many dreams are drabber, more incoherent, more splintered than waking life, dreams are put together from the same elements as our waking stream of consciousness and often verge on indistinguishability from it. Again, following Descartes' commonsense view, we think that the fundamental difference between our dream and our waking stream of consciousness is that, in the second case, real external physical mechanisms straightforwardly cause the objects presented in the stream to behave as they do.[2]

Roughly speaking, in my dreams I see and understand individuals much as I do when awake, with the same everyday assemblage of psychological notions that we use in sorting out our understanding of each other and of our own self. To construe others and myself in this way is to take up the *intentional stance* and to employ, as we do in narrative and everyday life, the *intentional idiom*. The point to note *here* about this age-old everyday stance and idiom is that it is, largely, dream/waking neutral because it makes no commitments to or suppositions about physical and neurological mechanisms. Hence, also, it is largely neutral between biography and fiction, stage and real life. In cunningly etched sentences Shakespeare has played, in tragedy and comedy, with these neutralities, but the graceful words Shakespeare gives Prospero – 'We are such stuff/As dreams are made on, and our little life/Is rounded with a sleep' – like the laceration with which he provides Macbeth – 'Life's but a walking shadow, a poor player,/That struts and frets his hour upon the stage/And then is heard no more' – echo on in the nursery's cautionary version of Neurath's metaphor

> Row, row, row your boat
> Gently down the stream
> Merrily, merrily, merrily,
> Life is but a dream.

But while the intentional idiom embraces these inversions and neutralities, it is inflexibly reciprocal and social. Whether we are in Birmingham or dreamland, in print, on film, or merely in the flesh, we are where we are *together*. Though Alan Turing, as we shall see in chapter eight, impishly refers to this refusal to entertain solipsism as 'a polite convention', the whole intentional idiom seems committed to it.

Prosaically, the intentional idiom gives us a *level of description*, and the intentional stance, a world of persons. Perhaps the most fundamental task of cognitive science is to discriminate between such levels: to tell the dance from the dancers (the message from the medium, the functional software from the hardware, to use more recent idioms). There are many other levels to worry about, as we shall soon see.

'Paradox' itself is an ambiguous, even paradoxical, word. Logicians mostly use it literally to refer to *unspeakable* sentences that, strange looping from truth to falsity and back, have all the potential for mental, rational, or mechanical self-destruction of evil beasts. But, of course, the discovery, taxonomy, and diagnosis of paradoxes has provided and will continue to provide us with insight and understanding, and a precise way for establishing the existence and nature of the *uncountable, unprovable, undecidable, unhaltable, and unmakeable*.

Further, 'paradox' is also used to mean a sentence, drawing, model, or narrative that initially or literally seems impossible but, taken more deeply or metaphorically, captures an important truth or insight, particularly when a commonplace fabric of relationships is turned inside out in some sort of Copernican revolution in perspective, as happens in the Shakespearian inversions I mentioned.

Our frontispiece, which replaces one of Escher's human hands with a mechanical one, *unbalances* the symmetry of the original. The analogous, interpretive text is from that inversionizer, Oscar Wilde, 'Nature imitates Art'.

The first and most obvious application of this inversion is this: while much of nature precedes artefacts temporally, and while the humans, literally, make the artefacts, *the priority and direction of understanding most often moves from artifice to nature*. After we build mechanical pumps, simply hoping to move liquid from a mine or a bilge, we may then turn about and understand what the heart is. Understanding pulleys, levers, and hoists historically proved the route through which we came to understand muscles, bones, joints, and sinews as the functional and architectural elements in arms and legs; but it is essence and generality as well, for the engineer's understanding is the more starkly basic, more easily grasped, rendition – *if we are to understand a limb we must first build levers*. The mechanical hand's purposeful structure is stark naked, while the human hand (though ultimately mechanical too) has been winnowed and chiseled out, grown and evolved, by the selfish genes, blindly conjuring up incredibly complex, densely compromised solutions whose architecture enigmatically recapitulates a myriad of previous solutions to other problems, each made from the architecture and materials of the one before.

Though the direction of understanding is from levers and pulleys to hands, the levers and pulleys were not designed to simulate hands, to look like or, still less, to replace them. Rather, levers and pulleys provided reach and force beyond that available to untooled men. On the other hand, artists and anatomists do have an interest in simulation. So, most revealingly, do *toymakers*, those paradoxical tricksters, machine punsters, whose work surely must fall under St Paul's interdiction, toymakers whose machines so fascinated Descartes that he fabricated such devices himself and argued that animals were machines, only to be followed, a century later, by Julien de La Mettrie, whose equal fascination led him to assert that humans, too, were machines. (In the twentieth century, engineers, with the need for hands to work controls in environments too dangerous for human flesh, naturally strove for more full-fledged simulations. We can be more serious about robots.)

The problem of building a simple mechanical hand is not an extraordinarily daunting one. So perhaps we could have understood our hands in such a way without first developing the lever/pulley technology. After all, nature, that slow but implacable artificer, invented the opposable thumb twice, as Stephen Gould shows us in *The Panda's Thumb* (the Panda's functional thumb is not a finger but developed as an enlargement of a wrist bone).[3]

But we have left something out, as has, of necessity, our frontispiece, though only literally, for metaphorically our frontispiece, like our cover, is a picture of the subject matter of this book. What we have left out is the control system. *That problem* is not at all easy or simple, or even simply mechanical, and, as I have said, it can only be pictured metaphorically. Even the problem of making the machine's hands grip delicately enough so that it can pick up an egg without breaking it is not trivial, though we have managed it, provided you do not include the problem of finding the egg. The problem of getting a machine to open the refrigerator door, pluck out the egg carton stuffed between the milk and the cheese and behind the bread, put the carton without mishap on the table and, finally, open it and pick out an egg, is at present beyond our grasp. And forget about making an omelet!

How do *we* do these things? We simply do not know. Indeed, we have only recently begun to plumb the depths of our ignorance and glimpse the dazzling complexity of our everyday cognitive activities. Though we have in most respects learned more about human cognition in the last 100 years, and particularly the last 40, than in the rest of our history, much of our achievement is in realizing just how much there is for us to be ignorant about. Accounts of future robots hatched by 1940s male science-fiction writers provide an ironic example. Isaac Asimov, Robert Heinlein, and hordes of other writers suggested that within a couple of decades we should have robots that did 'simple things' like house-cleaning, preparing and serving meals, and baby-sitting, while *of course* 'hard things' like astrogation and mathematics would remain for centuries, if not forever, beyond the powers of mere machines.

Through envisioning and building computers, and the input sensors and output mechanisms that they process and instruct, we have schooled our understanding of ourselves. We know more of the questions and, provisionally, some answers. But now a fearful question arises: how much of this is enhanced insight and understanding (of what we have been doing for a long time quite apart from computers) and how much invention, imposition, and transfiguration? Is our experience building formal systems and computers properly arming us to turn about and hew our cognitive nature at its supposed joints? Or are, and shall, we simply be hacking ourselves, in our theories or even in practice, into beings that make computer-technological sense? Is the computer a mirror or a mold?

The answer to this fearful question is, of course, both and neither, and all these in what we shall find to be a looping and fascinating variety of ways.

Looking at ourselves from the computer viewpoint, we cannot avoid seeing that natural language, in our present case, English, is our most important 'programming language'. This means that a vast portion of our knowledge and activity is, for us, best communicated and understood in a (and our) natural language. Even if we learn much more about the structural innards of natural language and about the processing levels and neurological realizations that manage its processing, we shall still benefit as well and learn as much when we read Dickens and Shakespeare, which, like the best of dreams and the most instructive of careers, enlarge our lives and understanding. Equally, Charles Darwin and Noam Chomsky, Bertrand Russell and Alan Turing, and their like in the future, will foreseeably primarily address us in natural language.

One could say that natural language was our first great original artefact and, since, as we increasingly realize, languages are machines, so natural language, with our brains to run it, was our primal invention of the universal computer. One could say this except for a sneaking suspicion that language is not something we invented but something we became, not something we constructed but something in which we created, and recreated, ourselves.

Paradoxically, for even puns have their place, *the medium is the message* ONLY if we are not interested in the message; and a good medium or language, as both human and the, much shorter, computer history make amply clear, minimizes noise and maximizes expressiveness, universality and informational integrity, reducing the significance of individual variation in hardware as much as possible. This only runs us in strange loops when the medium is what the message is about. But things are not as bad as Neurath's claustrophobic ship-builder metaphor might suggest. As recent work in linguistics shows us, we can in fact, in the manner of the natural scientist, stand aside and investigate human language objectively as an organic phenomenon, just as a physiologist can investigate the respiratory system. Natural language is such a robust phenomenon that it can be investigated without the elaborate precautions about subject or experimenter bias that psychological research requires when studying motivation, intention, attitude, and the like.

2

The Classical Agenda:
Plato's Problem, Aristotle's Turf-wars, Descartes' Solution

What is the final goal of studying intelligence: to build intelligent machines? to understand how the brain is put together? or to describe the structure and powers of intelligence as a free-floating entity, tied to neither brain or machine?

Hulbert and Poggio[1]

Plato's Problem

The term 'cognitive science' slipped into our vocabulary sometime early in the 1970s and 1980s, and interdisciplinary programmes with that title or the like are even more recent. Naturally, terms chase realities. However, even the fusion (or confusion) of questions, theories, inventions, and dreams that motivates the term is at most 50 years old.

None the less, in the fourth century BC, Plato asked two of the most central questions of cognitive science and gave the beginnings of an answer to both. Indeed, simply to put it that way understates what Plato achieved. He did not simply ask the questions. Rather, he showed why the questions had to be asked and why the most obvious, commonsense answers could not work. His own answers, and the evidence and argument he gave for them, have much in common with current views. Though he presented these views, at times, in metaphorical and mythic terms, we can claim him as the first substantial contributor to theoretical cognitive science. What I want to do in this chapter is not historical survey but rather to exploit a few brilliant anticipations of current work in a *preliminary* discussion of the major tensions in the question, 'what is cognitive science?'

Plato's two questions were:

1 Given the fuzzy and spotty sensory input that humans get, how do they come up with infinite, systematic, and abstract knowledge (as, for

example, in geometry, and mathematics more generally)? A major contemporary cognitive scientist, Noam Chomsky, calls this, 'Plato's problem'.

2 What are the kinds and characteristics of mental representations (ideas, thoughts, perceptions, opinions, and so on)? Do some and not others have a distinct clarity and infinite range of applicability and generality that sets them apart? (An affirmative answer to this last question seems to bring on 'Plato's problem', so the two are obviously connected.)

Plato raises and pursues these questions most strongly and compactly in the dialogue *Meno*. In it, Socrates questions a slave boy who has had no instruction in geometry. Socrates draws out of the boy a proof of a geometric theorem, even though the boy starts out by denying it. In Plato's presentation, Socrates, through a long series of questions, forces the boy to reason and to exploit his inner, tacit sense of line and angle (part of this is managed by having the boy draw shapes in the sand). The upshot is the claim that the boy *already tacitly knew the theorem*.

The boy's knowledge must be tacit in that he, at the beginning, denied the theorem. But Socrates does not proceed as if the boy had buried in him this one particular bit of knowledge, somehow suppressed, like the first name of a hated uncle. Rather, Socrates proceeds as if the boy had in him *a whole systematic way* of abstracting, reasoning, and generalizing, one that could have been encouraged to come up with any other geometric theorem as well. The tacit knowledge is systematic and constructive, like the sense in which those who know how to add know that $245+86 = 331$, though they have not memorized, or calculated, this particular equation. (Specifically, the Pythagorean theorem asserts that any right-angled triangle has the following property: the area given by adding together the squares of the two shorter sides is equal to the square of the long side. Socrates pushes the boy to see that certain claims *have to be true* and that from these, self-evident step by self-evident step, he must eventually realize that the theorem also is necessarily true.)

So Plato's answer to his second question is that one does have tacit systematic, abstract, and generalizing knowledge of what is necessarily true, whether one is an aristocratic philosopher or a slave or any other sort of human. As to the first question, Plato argues that this knowledge could not have arisen simply from repeated sensory experience, so it must be in some sense built into us from birth.

Up to this point many if not all cognitive scientists would strongly endorse Plato's whole line of thought. His questions are the right questions. His answers, so far as they go, are the right answers. His general argument for the built-in character of our tacit systematic knowledge of geometric form today bears the label 'the argument from the poverty of the stimuli'. Most cognitive scientists believe this sort of argument shows us how in general to distinguish the built-in structures of our knowledge from environmental input. The current term for built-in is 'native' or, for historically minded philosophers, 'innate', and the favored explanation for how we come to have such knowledge is that it is genetically programmed, as part of our normal cognitive maturation.

It is also true that there are a number of crucial ambiguities and unanswered questions in Plato's account. These have worried and subdivided subsequent work and constitute the agenda for much of cognitive science and, hence, for this book. Plato's student and critic, Aristotle, will provide us with some of the distinctions needed to pry apart ambiguities, sciences, and kinds of explanations. Before I go on with this, I need to make two points about Plato and his world.

The first is that Plato did, of course, have a metaphorical and mythic – and crazy – explanation of how Meno's slave boy and the rest of us happen to have tacit native knowledge of geometry and related matters. Plato's explanation is that our minds, before they acquired our bodies, 'flew about' and 'looked at' the pure geometric Forms (and also those of the True, the Good, and the Beautiful). What is crazy about this explanation is indicated by the quote marks in the previous sentence. How can you 'fly about' without wings or a body?; and what sense can we really make of nonphysical things having a spacial structure and position? How can you 'look at' something when you lack eyes or any other sensory equipment?

Someone might reply that it is all very simple, you just look within your mind with 'your mind's eye' and 'move about' in your imagination. We shall discuss the seductive dangers of this 'little man inside' talk subsequently. But even if it made perfectly good literal sense, this will not help Plato's explanation along at all. For him the Forms are the eternal objective structures that mathematical truths are about, whether these truths are apprehended by you or me or an octopodian Alpha Centurian or any other being that thinks, human or not. So Plato needs an account of what nonphysical seeing is and how it is to be distinguished from imagination. We shall return to this puzzle subsequently. Many have thought that Plato used such metaphorical and mythic talk when he wished to emphasize that a real explanation was required, though he did not have one.

The second point is an historical conjecture (one made by some modern historians, and one suggested by the ancient Greeks themselves).

Plato made the first substantial contribution to cognitive science as a member of the first human culture to engage in systematic rational inquiry. Why should the ancient Greeks have been the first scientists and so much else? Obviously, many factors contributed, and, since history happens only once, you cannot say what is essential with any assurance. But one bit of Greek technological innovation turns up as necessary in every modern account. With the slender addition of vowel letters to the various Semitic consonant systems, the Greeks formed a full-fledged written alphabet. This cognitive technology created an artificial memory, one more accurate and reliable than the natural sort, one in principle eternal and, most important of all, available to all readers of Greek.[2]

'The alphabet is not *technology*!' someone is bound to say. But of course it is. And it is a *cognitive* technological innovation with as revolutionary and as sweeping effects as any we know about. (Recent inventions of programming languages like BASIC and LISP, while allowing comparisons we shall consider later, surely are of much less significance.) For their part, the ancient Greeks most certainly regarded the alphabet as an explosive invention, one stolen from the Gods, along with fire, by Prometheus. Some of them, including Plato, worried about the distortion, falsity, and dehumanization that it introduced, much as current writers worry about computers and artificial intelligence. 'Compared to living, context-bound speech', so I would paraphrase this line of thought, 'written language is really just dead physical marks, an empty, fraudulent simulation of meaningful thought.'

We find similar arguments today, purportedly proving in principle that no computer, however many generations of improvements on present models there are, can ever really think or mean anything by its 'mere symbol manipulation'. Other ancient Greeks worried that with the written alphabet humans would soon no longer practise, and value, the memorizing of hundreds of lines of verse as a most central intellectual ability. They were right, just as those who have recently complained that cheap calculators have eroded our pride in school children who can perform routine arithmetical operations quickly and accurately – and our contempt for school children who cannot.

I need to add that while the ancient Greeks could wonder and worry about the invention of the written alphabet, they were much more unreservedly in awe of their native oral language. When Plato wrote his subversive *Meno*, slavery was accepted practice throughout the world and slaves, generally speaking, regarded as naturally suited to their role. Aristotle, for a revealing example, claimed that Persians and other non-Greeks were incapable of rationality and, therefore, were natural slaves. (Aristotle did allow that Greek women could absorb the form of reason from Greek men, though he maintained they were incapable of its active, creative employment.) Within this context, we can only marvel at Plato's audacity in suggesting that a slave boy could rediscover Greece's highest intellectual achievement in an hour or two.

When Socrates wants to show Meno that the knowledge of the Pythagorian theorem is innate, he asks not whether the young slave is male or Persian. He asks only whether the slave can speak Greek. I like to think of this conjoined with the passages in *Republic* where Plato has Socrates argue that women should be educated like men and that those who display the relevant qualities, as exhibited in the reasoned use of Greek, should be made rulers – all this in an Athenian Greece that did not allow women to participate in political, economic, or intellectual life outside the home! Tradition has it that the ancient Greeks invented the word 'barbarian', because they supposed non-Greeks were vocally restricted to mouthing 'bah, bah'. Though the Greek view that other languages were mindless noisemaking is surely wrong, I want to applaud Plato's claim that intelligent command of a human language demonstrates that one is a thinker and that one has much the same tacit knowledge and intellectual potential as any other speaker, whatever one's status, race, or sex (or, surely, species? – would not a computer in principle have some chance to qualify?).

Aristotle's Turf-wars

Plato conjured up what has to be the most extravagant and cosmic precursor of Wilde's epigram. As *Meno* suggests, Plato was impressed by the extraordinary success of the formal sciences, which alone had produced precise, necessary, and universal truths, and which did not depend on sensory observation or experiment. This led Plato to suspect that any search for knowledge should employ the same methodology, invoke the same standards, and take the same sort of formal objects as its subject matter. Consequently, Plato suggested that the so-called physical universe, the endlessly changing one perceived through our senses, was a shadowy simulation of formal reality. All of nature was an imitation of the artifice of eternity. All of science must be a unified search for the Forms that nature aped. In *Republic*, Plato even applies the geometric method to statecraft, producing at one point a 'proof' that the tyrant is precisely 256 times more unhappy than the just man.

To the contrary, Aristotle made a vast number of distinctions among the sciences in terms of goals, methods, subject matters, kinds of explanation, degrees of precision and certainty, and so on. We know that he systematically investigated everything from marine biology to meteorology, stagecraft to logic, physics to politics, and he headed the first research university (Democritus, more than a century before him, was a vigorous experimentalist with a materialist view of perception and cognition but his 60 books are lost). One wonders whether Aristotle's arsenal of distinctions was in part a product of his experience as a university administrator concerned to reduce conflicts between disciplines. In any case they will prove most useful for us in sorting out what kind of a discipline cognitive science ought to be, which existing sciences can properly contribute to it, and how all of it might fit together (if indeed it can).

In particular, Aristotle gives us an extensive analysis of the distinction between natural and artificial. This distinction (or the many distinctions conveyed by these terms) plays a central role in any discussion of cognitive science. But it is often taken for granted and that, I shall do my best to show you, is a mistake. Aristotle makes a related distinction between practical/productive sciences and theoretical sciences, one which would place computer science in the first group and cognitive psychology in the second, thus rather sharply raising the question of how both could possibly form a coherent science.

No thinker of our time can discuss such matters without the suspicion (and generally, the reality) of disciplinary bias. That includes me, of course, though I can hope that I am merely suspect and not guilty.

Today, computer scientists who do artificial intelligence often suppose their work is the real business of cognitive science, others having provided some initial formalizations or interesting data. 'Of course we fumble and charge up dead ends,' I imagine one saying,

Of course our strongest achievements have been in circumscribed areas, of course we cannot yet simulate the full range of a human's cognitive abilities, largely because we do not yet have much in the way of clear specifications of these multifarious but often shallow and limited abilities. But I shall tell you this: after only a few decades, our chess-players, theorem-provers, and medical diagnosticians, for example, *really* play chess, prove theorems, and diagnose illnesses, and do so better than almost any human, and we *really* understand how they do it, which is more than psychologists can say about any human cognitive capacity whatsoever after over a hundred years of concerted, systematic experimental investigation and over two thousand years of looser and more scattered study. The real truth in the Oscar Wilde quote is that you only really understand something *when you can build it.*

Equally, cognitive psychologists frequently see computer scientists as 'mere engineers' and unprincipled hackers who cannot do, or anyhow are not doing, genuine empirical research and who in any case do not much understand what they are trying to simulate. The familiar electronic computer has a hardware and software architecture that is not even remotely like the human brain and mind.

The AI fabricator is like a man who would disparage the psychophysiological study of human motor activity by building an automobile. His response to someone who points out that the automobile is not really much like the human bipedal version of the four-limbed mammalian architecture is that so long as you have a well-paved road, his automobile can reach a higher speed; and when you point out that his model does not include a driver, he waves his hand hopefully but emptily down the road.

I need hardly add that both the AI computer scientist and cognitive psychologist, as I have caricatured them, would find perhaps their only agreement in the following thesis: that philosophy, logic, mathematics, and linguistics, because they neither build cognizers nor empirically investigate the cognizers we have already got, have no proprietary interest in cognitive science.

As my disciplinary affiliation is logic and philosophy, I hope you will not expect me to caricaturize the exclusionary and prideful claims of philosophy as I did AI and psychology. Why not assume them to be arrogant, sweeping, and bogus? You will have some examples of this later on when we shall review some philosophers' dismissals of machine intelligence.

But though Aristotle is called a philosopher, do not tar him with the same brush as me. He was not a philosopher if by that you mean someone who would today bear that disciplinary label as opposed to a physicist, biologist, psychologist, mathematician, and so on. It is absolutely clear that he carried on *all* these roles and that the word 'philosopher' in his time meant no more one than the other – it meant them all and so might best be translated as 'scientist'. So I hope to present Aristotle's distinctions as unbiased by a particular disciplinary commitment, as an analysis of common sense and notions more general than particular sciences, and as an attempt to conciliate the claims of particular scientific disciplines.

(A classicist philosophy professor might complain that I fail to mention that 'philosopher' literally means, and was so coined by Socrates to mean, 'lover of wisdom'. Frankly, I simply have not noticed a more zealous lust for the stuff among philosophy professors than among others. What stops this etymological quibble is this. Plato's presentation of Socrates' point, in coining the term, was that Socrates would not accept any payment for his instruction, as opposed to the more professional sophists, who took tutorial payments. All professors of philosophy who on principle refuse to accept payment for their professional activity may now stand up.)

I mentioned that Aristotle distinguished theoretical and practical/productive sciences, the first pursuing knowledge, the second happiness. Theoretical sciences aim to describe and explain the universe. Naturally, there are as many such sciences as there are subjects to study (physics, biology, and mathematics, to use some of Aristotle's examples).

Master practical sciences such as politics and ethics aim at successful decision-making, at the communal and individual level, more than theoretical understanding. Similarly, the more exact but subordinate productive sciences such as medicine, engineering, navigation, and agriculture aim at health, sound buildings, reliable landfalls, and successful growth of food. (These productive sciences are subordinate because who shall be cured, what buildings built, destinations approached, and crops raised are all political and ethical decisions. Aristotle's notion of political science is not that of describing and explaining the human communal decision-making structure in the way a zoologist might characterize the structure of an ant community. Rather it is like medicine or engineering as opposed to physiology and physics: the good political scientist should cure states, as the physician cures sickness.)

Is cognitive science, then, supposed to be a theoretical or a productive science? Is it like physics or engineering, chemistry or pharmacy, geophysics or navigation, zoology or animal husbandry?

The questions of the last paragraph may suggest a tilt toward cognitive psychology as opposed to computer science. Surely, if both sciences are concerned with the same natural phenomena, but one with investigating and explaining them and the other with putting this knowledge to use along with a grab-bag of technological tricks, we may think of the theoretical science as basic, the productive one as derivative.

However, we may do well to reflect on a peculiarity of the cognitive 'behavioral sciences', one that calls into question their credentials as theoretical sciences and hence also the derivative status of their productive counterparts. Take, for example, sociology, economics, and psychology itself. Abstractly, one would think that sociology ought to be the science of interactions and community between entities with social interests and substantial cognitive powers. Hence, it would seem it ought to be a science with no principled interest in human interaction as opposed, perhaps, to that of intelligent nonhuman Earth animals *and* possible extraterrestrial entities with cognitive social powers comparable to or bizarrely different from ours. You can see that the same kind of point can be made about psychology and economics.

It was once thought that physics and chemistry were concerned with earthly phenomena alone, since matter outside our atmosphere could be assumed to operate on completely different principles. But every physicist or chemist today assumes that his subject matter is the universe, not the peculiarities of Earth, which are, though in the past misleading in practice, presumably in principle no more telling than those of Mars or the Magellanic Clouds. There may be no stars, quasars, black holes, or comets on Earth (thank God) but a physicist would think you a crazed medieval relict if you denied that they are just as central a part of what physics has to explain as anything that goes on here on Earth.

Similarly, chemists did not expect, or get, any big surprises when we brought back sample moon rocks or radioed back data on Mars' surface, for they felt, and had a right to do so, that their science already knew what had to be the case on the Moon and Mars. *If* there had been some really astounding surprise such as a new element – not a previously unobserved but theoretically predicted element way high up on the periodic table like californium or plutonium before we manufactured them, but an element incompatible with the table, one with, say, nine electrons in its penultimately inmost ring rather than the standard budget of eight – this would *not* mean a new science, 'Martian alchemy' or whatever, standing alongside but completely incompatible with regular chemistry. No, it would mean current chemical theory contained some huge basic error. Chemical theory would have to go through some revolutionary reformulation in which the old periodic table might re-emerge as a special case, along with the nine-ringer, of an extraordinarily different chemical theory (this is sometimes said to have happened to Newtonian physics within the wholly different theoretical structure of Einsteinian physics).

On the model of physics and chemistry, is it possible to hope for a nonprovincial cognitive science, one concerned with the cognitively possible, impossible, and necessary on a universal scale, not with the mere naturalistic description of one very particular biological species on one planet among millions that have conditions favorable to life? Given the excuse of comparative immaturity, should not sociology and psychology none the less assume that in principle their subject matter is socializing and thinking entities everywhere, not just on Earth or within the human species?

The point of this question, of course, is that physicists and chemists present us with general theories of what must and cannot happen physically and chemically throughout the universe. Their experiments, almost entirely conducted on Earth for obvious practical reasons, are structured towards this more universal concern. If physicists and chemists were asked to confine themselves to a mere episodic description of what actually happens here on Earth, without any concern with what can or cannot happen (for this would involve a theory of physical and chemical possibility and necessity that would make commitments far beyond any earthly contingencies), I think all respectable physicists and chemists would consider their occupation had come to an end. Should psychologists and sociologists be content with less? Do some inwardly feel that they simply have no hope of offering general theories and must content themselves with offering nothing more than an impressionistic historical account of some social and cognitive activities of a particular biological organism on a particular planet during a short time period? Even if one claims humbly to describe simply what happens (like a newspaper reporter or a narrative historian), you cannot really hope to do so, for what happens is textured and contoured by what can and cannot happen. Worse, if these studies have no principled universality, then their productive counterparts are *not* mere derivative applications and imitations. The relationship may begin to look more like that between historical accounts of human shelter-making and the productive science of architecture.

Psychologists, economists, and others who study human cognitive activity have three reasonable replies to this criticism of their pretention to doing theoretical science.

The first is that these are fledgling sciences with complicated subject matters. Hence we should expect description and not too much theory. Second, the nonhuman animals of Earth, who lack consciousness, language, etc., are so limited and unlike us cognitively that study of them really is another undertaking. And while it would be rash to deny the existence of powerful extraterrestrial intelligences, we have absolutely no data about these hypothetical beings (while telescopes and the like provide us with a lot of information about extraterrestrial physical and chemical phenomena). In the highly unlikely event that we did get in contact with such intelligences, their cognitive processes might well be totally unintelligible to us. Though the chimpanzee, for example, is almost indistinguishable from us genetically, we have virtually no idea of what its inner life is like. How incalculably more difficult it would be with aliens who undoubtedly would not even have the double helix DNA genetic material that we share with every living organism on Earth. Thirdly, we do have some theories, or idealizations, in psychology, economics, and elsewhere that have generality and may well universally apply to rational beings, not just humans.

I shall make some brief initial comments on these responses in the next three paragraphs, though we shall return to these issues throughout the book.

Psychology is a young science if by it one means the experimental psychology of the past few decades that prides itself in doing experiments with the sort of elaborate controls and enormous statistical sophistication that you rarely find in the natural sciences. Economics and sociology, were you to do it this way, are equally recent. But if you consider the models of rationality, of more general than human application, mentioned in the third response, psychology is ancient indeed. Aristotle put together a model of rationality – logic – that lasted until the middle of the nineteenth century. His logic, with some emendations, was eventually incorporated in a more general model by Bertrand Russell in his *Principia Mathematica*, a turn-of-the-century milestone in cognitive science. Can we regard Aristotle and Russell as having made contributions to universal cognitive science, characterizing rational beings in general?

What about mathematics? Some people are impressed by the suggestion, sketched above, that since the chimpanzee, nearly identical to us genetically, differs greatly from us cognitively, it is likely to follow that aliens who differ greatly biologically must be still more unintelligible. To the contrary, astronomers who speculate about how to formulate a message to send to the stars, a self-contained one that could be deciphered by aliens who might receive the message a thousand years in the future, invariably assume that the aliens will have the same basic logic and mathematics as we and in particular will recognize the same prime numbers as we do (knowledge of the primes will be the key to decoding the first part of the message, on which the rest of it will piggyback, using of course logic all the way through). These same astronomers also assume, less centrally, that alien intelligences will know much the same physics and chemistry as we. Is it possible that when intelligence takes off, reflectively generalizes itself so to speak, it will converge on certain kinds of knowledge, no matter what the biological or chemical hardware that supports cognition? (Mathematicians, logicians, and natural scientists are very prone to make this assumption, while many behavioral scientists and philosophers find it bizarre.)

Aristotle claimed that there are four different basic kinds of explanation. Leaving aside material explanation, the characteristic feature of something natural is that ancestral, purposive, and structural explanations converge, whereas all three may be quite separate for artificial objects. An oak tree, for example, is parented by oak trees and it is formed so as to produce still more oak trees. On the other hand, an oak wood chair's purpose is to seat a human, though chairs do not grow on trees and will not, if planted, multiply. As he puts the distinction more generally, a natural object has its source of motion within, while for an artificial object, it comes from without.

Two points need to be made about Aristotle's sensitive analysis of the natural/artificial distinction. The first is that the distinction loses much in significance in mature sciences like physics and chemistry. The second is that the electronic computer can come to be understood to be natural.

Several paragraphs back I mentioned that the chemical elements, plutonium and californium, were predicted before they happened to be manufactured – the periodic table had a place for these elements and their properties were surmised. This had happened earlier in the history of the table, when predicted elements were eventually found occuring naturally. But chemists cannot make any *chemical* distinction between a pure sample of an element that has been artificially produced and one that has arisen naturally. (Anyone who has looked into the maze of considerations involved in distinguishing 'natural foods' from ones 'full of artificial chemicals' realizes how tenuous the distinction can be. Why not take a clear line and argue that no food is natural unless it grew without any human intervention or culturing? How much sense does it make to say that human intelligence is 'natural', given artificial technologies like the alphabet and all the others that have followed in its wake?)

Among living creatures we can make a distinction between those that we have deliberately bred for traits that interest us and those that have been, as Darwin long ago observed, bred by the pressures of the environment. Here Aristotle's distinction has some purchase. The chicken and the pig are artificial, the hawk and the wild boar are natural, in that the structure of the chicken and the pig have to be explained in part in terms of what humans have done to and planned for them. More direct genetic alteration is now becoming possible and this will underwrite another sort of artificial/natural distinction. None the less, with our growing knowledge of biochemistry, genetic structure, and cellular development, biology has moved in the direction of physics and chemistry. By far the greater portion of chicken and pig development and structure is now explicable on the same biological grounds as that of the hawk and the wild boar – we have but nudged nature a little. Much the same will be true for anything that gene alteration is likely to bring.

Matters of course were not so some centuries ago, when life was regarded as a mysterious, nonmaterial force, or demi-soul. Today school children at the Smithsonian Institute in Washington DC watch the cooking teacher, Julia Child, on video tape, making 'organic cosmic soup', a demonstration that simple inorganic chemicals, exposed to radiation and electricity, will form simple amino acids. Yet not so long ago the artificial or mechanical attempt to create life was regarded with horror, as some sort of unnatural, God-challenging blasphemy; and the result of the attempt had to be, not genuine life, but 'the undead', some kind of gruesome simulation that, lacking the legitimizing demi-soul, was necessarily evil. I think that something like the same attitude is present today towards artificial intelligence efforts.

Descartes' Solution

In the seventeenth century, René Descartes incurred the accusation of life-simulating blasphemy during the course of his general attempt to solve the problems bequeathed to him by Plato and Aristotle (without much significant intervening progress). Of a medical family and inspired by the new mechanical physics and recent technology, he strove throughout his life to find mechanical and physical explanations for life, for the behavior and inner structure of animals (he was also greatly concerned that medical cures would stem from such research). Often termed the founder of physiology, Descartes proceeded both as experimenter and engineer, dissecting and occasionally vivisecting, but also building simple mechanical models of skeletal and muscular function.

Indeed, Descartes was accused of atheism and besmirching the spiritual mystery of God's beautiful creation because he claimed that the activity of biological organisms could be explained entirely in a mechanistic way (the 'beast-machine' hypothesis), thus dispelling the 'animal-soul' view that had been commonplace since Aristotle. (Though Descartes' position is generally accepted today, his supporters lost the battle in the 50 years after his death and, under the label '*élan vital*', the view that life required some special nonmaterial spark persisted into the nineteenth century.)

In part, Descartes appeared to argue that the activity of animals (and of humans in so far as they are animals) could be explained in a wholly mechanistic way *because* he wanted to contrast this bodily aspect of the world and of human beings to something else. This something else that demonstrably could *not* be explained in this way, namely, mind, would be the subject matter of a cognitive science whose laws would differ radically in explanatory character and level of abstraction from those of physical science.

Investigating Plato's problem with considerably more rigor, Descartes argued that what we find in our mind seems to be a mixture of active native formative forces and passive environmental sensory inputs, with some items relatively pure cases of one or the other. In *Meditations*, Descartes gives an example of both:

1 a thousand-sided regular polygon as a 'clear and distinct idea' (as understood by someone alert to its nature, as Meno's slave boy has become alert to what must be the relationship between the hypotenuse and the other two sides of a right-angled triangle);
2 a thousand-sided regular polygon as a mere 'adventitious image', something passively taken in by the senses as when we might look at what

we were assured was such a figure or remembered such an experience (as when the slave boy first looked at the triangle drawn in Athenian sand).

In the case of the idea, Descartes points out, we can establish a number of necessary points (such as that there is a side directly opposite, and parallel with, any particular side – and precisely 499 sides between either to the left or right). In the case of the image, any attempt to count the sides, to assure ourselves that the sensory image we are looking at is in fact a thousand-sided figure, is way beyond our imaging capacity. In this example, Descartes tries to sever the native and adventitious elements that normally fuse in ordinary perception. More typically, our mental representations in ordinary perception of the world around us natively structure sensory input into representations of three-dimensional objects. Descartes insists, and very recent research into newborns confirms, that the notion of substance, of some objective unity underwriting visual, auditory, and tactile experience, is indeed innate. Our thoughts are not, as some empiricist philosophers have said, decayed sense at whatever stage the rot has reached, rather we find them natively structured at the particular levels that our minds are natively prepared to process.

Descartes summarized his distinction between cases where mechanical explanation sufficed and where native mental explanation was required in a test. Seeking the limit to mechanical explanation, he pointed out that it would not be difficult to build a robot that sounded 'I am in pain' when you pressed it and so on (my supermarket checkout counter possesses this capacity and more).

What *would be impossible mechanically*, so Descartes thought, was to make a robot that 'would reply appropriately to whatever was said in its presence'. The reason Descartes thought this would be impossible was that such a robot would have to have *abstract capacities for discriminating an infinity of possible inputs and appropriately generating an infinity of responses* in order to model the world and reason about it. Such abstract capacities were for Descartes the touchstone of the mental as opposed to the mechanical and physical. (While no computer robot today can 'reply appropriately to whatever is said in its presence', we lack Descartes' confidence that one never will. Indeed, Alan Turing, whom I propose as the first full-fledged cognitive scientist, proposed essentially the same benchmark in 1950 and it is now called, 'the Turing test'.)

To this point, Descartes' line is one that has been in command, though not without criticism, over the last 30 years of cognitive science. But Descartes had a further doctrine, metaphysical dualism, that captivated nearly 300 years of philosophical discussion and is today nearly universally rejected, though much of what led him to this doctrine is still part of the budget of cognitive science.

Descartes' famous sentence, 'I think; therefore I am', stakes the claim that the conscious mind is an independent substance, totally different from the mechanical body and capable of independent existence. To his familiar emphasis on a diversity, and natural hierarchy, of mental representations, from passive and sensory to actively structural, Descartes here added the argument that he could be absolutely sure, while thinking, that he as a thinker existed, while at the same time he could doubt whether he had a body.

This possibility of doubt about the physical world so came to impress subsequent thought that the dominant view for two centuries was that it could never rightly be overcome. Similarly, many were so convinced by Descartes' claim for the independence of mind that they doubted whether any sensory or bodily evidence could convince them of the existence of minds other than their own. After all, as Descartes pointed out, the mind cannot be perceived by any of the senses but only experienced in thought from within. Hence, it has no shape or dimensions as physical objects do and it is in this way an indivisible unity. While many were so led to a kind of solipsism about other minds, Descartes himself accepted the test (now the Turing test) mentioned above.

Aside from general distaste for mechanistic explanations of life, which had decisive effects in his century, two problems vitiated Cartesian dualism. One is simply that the formal means for making his case for nativism and rationalist mental representations sufficiently sharp and conclusive were not available, only fully appearing in the twentieth century. The other is that metaphysical dualism, the claim that the mind is a wholly independent substance of an entirely different nature from the physical body and brain, simply poses insoluble problems. If the mind is so wholly different and independent, how can what obviously happens – interaction between body and mind – happen? Though following centuries were to see some truly extraordinary proposals for a solution to this problem in metaphysical dualism, none would carry the day.

3

The Gathering Storm:
La Mettrie's Machine, Frankenstein's Monster, Babbage's Engines, Russell's Logic

Man is to the ape, and to most intelligent animals, as the planetary pendulum of Huyghens is to a watch of Julien Leroy. More instruments, more wheels and more springs were necessary to mark the movements of the planets than to mark or strike the hours; and Vaucanson, who needed more skill for making his flute player than for making his duck, would have needed still more to make a talking man, a mechanism no longer to be regarded as impossible, especially in the hands of another Prometheus . . . . To be a machine, to feel, to think, to know how to distinguish good from bad, as well as blue from yellow, and to be but an animal, are therefore hardly contradictory. I believe that thought is so little incompatible with organized matter, that it seems to be one of its properties on a par with electricity, the faculty of motion, penetrability, extension, etc.

J. O. de la Mettrie (1747)[1]

La Mettrie's Machine

I mentioned that Descartes denied the existence of an animal's soul, maintaining that animal activity (indeed, much of human activity) could be explained in a mechanistic way. Hearing this, you might have suspected that opposition to Descartes' views came not from liberationist sympathy for animals but rather from the simple thought that if we allowed that animals might be machines, the next step would be to make the same claim about humans. The suspicion is reasonable. Descartes' time abounded in shameless, intentional, wholly uncriticized cruelty to animals, arranged animal fights constituted a major form of amusement, a stern test for bull and bear dogs (modern versions of Grimms' fairy tales mute this, changing for example the boyish hero's amused, casual killing of birds to flies in 'Seven at a Blow'). Descartes' critics complained just as much that he offered mechanistic explanations of the sun and the movements of the planets, and their claim that animals had souls did not, for most of them, imply any need to change the treatment of animals, though it did imply an end to scientific investigation of them.

It was then perhaps inevitable that someone would propose that the entire human being, cognitive aspects included, is a machine. That is what Julien Offroy de la Mettrie (1709–51) did, publishing in Paris in 1745 *The Natural History of the Soul*. While the book attacked dualism, La Mettrie, surely to avoid censure, glossed a version of Lockean empiricism with a pseudo-Scholastic terminology, even suggesting that his account was ultimately Aristotelian in character. His Lockean antimaterialist dodge was not successful; nor anonymous publication with the claim that the book was a translation from an imaginary English work by a Mr Sharp. The Parlement de Paris had the book burned by the public hangman and La Mettrie fled to Holland where he published, in 1747, the first straightforwardly mechanistic and computational account of the human mind in his powerful and popularly written *Man a Machine*; here he defends a Cartesian mechanistic account of animals and extends it to humans, even saucily suggesting that Descartes himself had held the same position but concealed it for fear of persecution. As a practicing physician, La Mettrie suggested that it was but obvious common sense that, since brain injuries had obvious and selective psychological effects, one's mind could be nothing other than 'one of the properties of organized matter': to the challenging question 'How could a machine think?', La Mettrie's countering paradoxical inversion is 'What else could?'

The Dutch republicans were no more tolerant than the cosmopolitan French. A completely nameless authorship only temporarily saved him, though he had the curious satisfaction of writing of the antimechanist persona he had protectively assumed, 'Let the so-called M. Sharp mock philosophers who regard animals as machine; how different my own view!'

His identity discovered, La Mettrie fled. Fortunately, he received protection, honours, and a court position from Frederick the Great, spending his last three years in Berlin. There he wrote medical studies on asthma and, possibly ominously, on dysentery. In *L'homme plante* he argued for a continuity between all biological organisms, and in *Les animaux plus que machines* he satirizes the view that animals contain some spark (or soul) that distinguishes them (and humans) from the rest of nature. He also wrote *A Discourse on Happiness*, and *The Art of Pleasure*, which took a physiological view of happiness and breezily commended sexual as well as intellectual pleasure, indicted religious enthusiasm, defended atheism, and condemned remorse as a pathology.

La Mettrie was viciously slandered after his death, even by those who adopted his ideas without acknowledgement ('evil beast, slow belly', as St Paul would say). While for several years *Man a Machine* was widely available in Prussia and England, and in France too though unofficially, little serious discussion of his ideas could survive the scandal and slander that engulfed him; those who attacked him seemed to feel that even mention of his name or works, or exposition of his actual views, would pollute their readers. Hence, La Mettrie become 'one of the most abused, but one of the least read, authors in the history of literature', as Professor F. A. Lange wrote in *The History of Materialism* in 1865, a work that finally gave a somewhat more respectful treatment of La Mettrie and established his priority on many issues. None the less the good Victorian Professor Lange found La Mettrie's assertion (or mention, perhaps) of the importance of sexual pleasure abhorrent and, to note what seems to be his most fundamental general criticism of him, *frivolous*:

In common with most of these French philosophers, he found only people who loved the debaucheries of sensuality as much as he did, and who only took care not to write books about it. La Mettrie may be frivolous, and this is a serious charge, but he neither sent his children to the Foundling like Rousseau, nor betrayed two girls, like Swift; he was not convicted of corruption, like Bacon, nor does the suspicion of forgery rest upon his name as upon Voltaire's.

(In 1925, Bertrand Russell introduced the English translation of Lange's book from which I quote; indeed, Russell there defended Mettrie's claim that the man–machine view was a natural extension of Descartes' views. In our own time Russell has suffered, if not the oblivion that Mettrie's work endured, an analogous experience, for in 1940 Russell was barred by a New York City judge from teaching logic to City College graduate students because, among other instances of what Lange would call shameless frivolity, Russell had written in *Marriage and Morals* that it might be a healthy idea for parents occasionally to appear in the nude before their young lest children acquire the notion that the human body is inherently obscene. Though Russell is often termed the greatest philosopher of our century, no American university or college was then willing to risk offering him a job, and, indeed, serious discussion of his work waned through the 1940s and 1950s, though brisk, and now discredited, 'refutations' of his logical and linguistic theories were common. Though professors of philosophy often enthusiastically quote the philosopher Socrates' characterization of his profession as that of a gadfly on the body of the state, they paradoxically but all too understandably condemned Russell as unprofessional for maintaining the Socratic role.)

While Professor Lange correctly insisted that in all the eagerness to condemn La Mettrie no specific accusation of any immoral behaviour occurred, Lange blamed him for the absurd canard that was used to condemn him after death – that he expired in gluttonous ecstasy immediately after consuming an enormous quantity of pâté laced with truffles (while the King James translation has 'slow belly', more recent translations give us 'swollen belly' and, most revealing but least literal, 'lazy glutton'). The account is still echoed in the current *Encyclopaedia Britannica* which states 'a carefree hedonist to the end, he died of ptomaine poisoning'. Aside from the fact that ptomaines, first named and described in 1875, were soon discovered to be harmless, so that the unlearned term 'ptomaine poisoning' can only mean some sort of chemical poisoning (other than ptomaine), the account of the one eye witness and of Frederick, in his 'Eulogy on La Mettrie' and in his letters, are incompatible with food poisoning and strongly suggest that he died of an infectious disease and a dangerous experimental cure that La Mettrie, in high fever and delirium, insisted on testing on himself. In any case, La Mettrie would have come upon the pâté in question as a physician who had attended and apparently cured the French Ambassador to Berlin, Tyconnel, of some grave illness (even possibly the one that would infect him); with the late hour, he was impulsively invited to supper by the Ambassador's wife, where he may have eaten the pâté or what Frederick described as a pastie. Shortly after the meal, he was overcome by high fever and delirium. This, and the 20 days of agony that followed before death, is hardly 'carefree until the end'. Frederick, while correctly emphasizing that La Mettrie, at 40, might well have corrected and expanded his hurried writings, publically eulogized La Mettrie.

He boldly bore the torch of experience into the night of metaphysics; he tried to explain by the aid of anatomy the thin texture of understanding, and he found only mechanism where others had supposed an essence superior to matter. . . . Nature had made him an orator and a philosopher; but yet a more precious gift which he received from her, was a pure spirit and an obliging heart. All those who are not imposed upon by the pious insults of the theologians mourn in La Mettrie a good man and a wise physician.

The King of Prussia was La Mettrie's sole defender. To this very day, the only work of La Mettrie available in English translation is *Man a Machine* and that only in a clumsy and inaccurate 1912 scholarly translation; the sophisticated Princeton University Press edition of 1960 gives no translation, though the editor allows himself 160 pages to comment in English on the slender 60-page French text. Professor Noam Chomsky has labeled 'Orwell's Problem' the problem of how an ostensively free society can 'bury' an indigestible but all too obvious line of thought and, perhaps as well, the individual who proposes it. La Mettrie, and perhaps

Russell and Chomsky himself provide instances. Cognitive science is an exciting but also, clearly, a dangerous occupation.

Frankenstein's Monster

The clockwork, mechanical man, as indeed the clockwork solar system and universe, fascinated intellectuals and appalled churchmen. There were echoes of this at the level of popular culture both in mechanical puppets and toys, and in tales stretching back to the golum of medieval Jewish legend and beyond to the moving statues of classical antiquity. Yet strictly mechanical devices were long familiar and, in their risible ticking and clanking, and jerky motion, apparently quite unlike living things (Descartes' strict notion of mechanism already clearly required the addition of chemical processes to explain organisms).

It took the very suggestive discovery that electricity was the same force in lightning, in the nerve impulses of living organism, and in the just developing technology of batteries and generation, to reanimate popular fears of the ungodly creation of sentient life. For Mary Shelley, a young woman of 21, to have written an instant bestseller, *Frankenstein* (1818), suggests that the culture of the early nineteenth century had built up a considerable charge, waiting for this conduit among others.

By something like the comforting illogic that leads parents to transform children's fears of cancer and atomic war into witches and bogeymen, Hollywood medievalized Mary Shelley's ambiguous vision. While she imagined a brilliant, young graduate student, studying physiology in a most modern Swiss university town, electrochemically creating life from inorganic materials, the 1930s film version brings us the brooding outcast in his Transylvanianesque castle, sewing together grave-robbed body parts. Hollywood's monster has only a little of the limited intelligence of his grave-gotten brain, with hardly any grasp of his situation and a most savage temperament.

Shelley, however, imagined a truly new-made man and so had to explain how this artificial man learned language and how he attempted, with great intelligence and initial compassion, to understand the world around him and his own horrifying situation. We are told that she began the work when dared to write a horror story, along with Lord Byron and Percy Bysshe Shelley, who did not complete their own attempts. What is extraordinary is that, as the story expanded beyond its original frame, she felt forced into a sympathetic account of the monster's cognitive and moral development from his own point of view. Her mother had written

A *Vindication of the Rights of Woman* (1792), possibly the first full-scale demand for equal rights for women.

Shelley imagines her monster learning language by hearing and seeing a blind, fugitive father, and his daughter, teach French to the romantic young man who had rescued her from some outrageous assault. Knowing his appearance is hideous to the prejudiced human eye, the monster waits until the blind father is alone, hoping to explain himself in language and seek help and understanding. Alas, the daughter and lover return before he can exchange more than a couple of anxious sentences with the blind father, and the monster is driven away. After some other attempts to aid humans that end in a similar way, the monster seeks out his maker to demand that Frankenstein create him a mate, that he may have some companionship and live apart. After initial agreement, Frankenstein refuses and the monster metes out much in the way of revenge but never so much that Shelley allows us not to feel there is some terrible justice in his actions.

The tale is high romanticism indeed but if we need a myth for our times, Mary Shelley's one, aptly subtitled, echoing La Mettrie, 'The New Prometheus', is not unsuitable. Shelley's vision also makes clear the centrality of language, as opposed to physical appearance and nonlinguistic behavior, to cognitive personhood. The monster's only hope is linguistic cognition and communication. It is that ability that inevitably leads the reader to acknowledge him as a person.

At much the same time as Hollywood, the adolescent Alan Turing read Shelley's book with interest. When he set out on his self-assigned project of 'building a brain' in 1944, one of his earliest, light-hearted proposals, as to what one should do with a compact computer of human-scale intelligence, was to pop it into a robot and set it off walking in the countryside (just as Shelley's monster had been so inadvertently released). Sometime later one might interview it on what it thought of humans and their world. His rather more guarded 1950 essay, 'Computer Machinery and Intelligence', harps on the notion of 'playing fair with the machine' in a way that still seems to most today to be at best facetious and at worst treasonous to the human race. (It is instructive to learn that the word 'robot' was explicitly derived from a Czech word meaning 'slave'.)

Babbage's Engines

It is startlingly agreeable with Percy Bysshe Shelley's notion of the romantic poets as legislators to the world that the teenage runaway who eventually became his wife should have written the myth of our age. It is astonishing that Lord Byron's daughter, Ada Lovelace, should now be generally regarded as 'the first computer programmer'. Beyond that, it is astounding that the British government, leaning on the advice of the Duke of Wellington, should have, beginning in the 1823, made what is proportionally one of the largest individual research grants in its history to Sir Charles Babbage to build what were eventually to become preliminaries for the world's first computer, the 'Analytical Engine' (completely envisioned by 1834 and partially constructed, the Analytic Engine had Babbage's young friend, Countess Ada Lovelace, as its first programmer and proselytizer[2]).

The Engine made perfectly good theoretical sense, and was a full-fledged computer in that it was able to combine arithmetical processes with decisions based on its own computations. It had an internal memory and was fed by three sorts of punch cards: programming cards, elementary arithmetic operations cards, and data cards (much like the familiar punch cards with which Herman Hollerith mechanized the US census of 1890 before going on to found IBM). The Engine also realized the notion of subroutine, automatically printed out its operations, and could modify its own programs. Unfortunately, the extraordinary precision required in the extraordinary number of gears that the Engine needed for its internal memory and processing, and the practical difficulties that the energetic but impatient Babbage encountered, meant failure in practice.

In a manner suggestive of more recent work, Babbage began with the clearly feasible but delicate task of building the 'Difference Engine' to calculate logarithms more quickly and reliably. When problems occurred, some involving disputes with his master mechanic (who at one point encumbered several rotors), he found it too easy to go on to the Analytic Engine, which would do the logs as just a small part of its general and powerful computational budget. But this Engine, of course, required even more precise and extensive engineering, and much more money. Even with the use of much of Babbage's own fortune, the work eventually stalled and, by the mid-nineteenth century, the effort was dead, not to be relaunched for nearly 100 years.

Some of my colleagues like to argue that, though indeed Babbage's Engine is theoretically possible in a world without friction, heat, and impossibly perfect gear grinding, none the less it is practically impossible in the real world for just such reasons. I feel a partly emotional dissatisfaction with this view. Similarly, I am quite charmed that fairly early in the history of the Massachusetts Institute of Technology (MIT)'s efforts in artificial intelligence a device for playing an unbeatable game of tick-tack-toe was built out of wooden tinker toy parts. It worked, though breakdowns happened. Despite the notable cachet of electricity, which

Mary Shelley evoked, it is as well to remind ourselves that theoretical computational capacity is what is essential in principle, not reliability or speed, or, still less, the materials employed. (This point will be more richly apparent in the presentation of the Turing machine in the next chapter. I do not mean of course to disparage the enormous *practical* importance of the inventions such as the transistor, for example, in 1948 and latterly, the microchip; the first electronic, vacuum tube, computers, which the transistor made obsolete, were enormous, unreliable, and very limited in capacity.)

Russell's Logic

This chapter has proceeded so far by three short, jerky steps. La Mettrie simply raises the question, naturally following out of Descartes' approach, as to whether mechanistic or physicalist explanations can cover human activity as a whole. But the formal means for evaluating his proposal were lacking, so the public hangman's bonfire is simply a slightly cruder reaction than others that were possible. Shelley provides a spirited and ominous myth but certainly little more. With Babbage we have a restless engineer who envisioned, and took a number of practical steps to build, a machine that would perform any combination of (finite) mathematical calculations reliably and quickly. If we are to trust Lady Lovelace's pseudonymous remarks, Babbage's view was that the Analytical Engine could do nothing other than to number-crunch as it was generally programmed to do: it could not be thought to think.

Perhaps Babbage presents an ideal case of something whose success is now familiar in computer science (a kind of hardware variety of hacking): someone managing by main force and gimmicks to simulate a cognitive capacity that is not sufficiently formally understood. For until an accumulation of work that eventuated in Alan Turing's 1936 paper, 'On Computable Numbers', we had no formal analysis of what 'any combination of (finite) mathematical calculations [i.e., computable functions]' might mean. Even more substantially, Babbage did not have available a logical language, or an adequate stock of logical and computational abstractions, with which to describe abstractly or in mechanical detail what his machine might be doing or what it might be asked to do. We may want to agree with the computer engineer's version of Wilde's epigram: you can only understand something complicated by building it. But we may want to add: abstract art(ifice) may have to precede concrete engineering.

Now comes the formal understanding that will underwrite the later, successful effort. It might, in the spirit of the last paragraph, be called 'abstract cognitive engineering', but in fact it is referred to as logic, foundations of mathematics, and computability theory. (Similarly, computer science, even more clearly, ought to be called 'computational engineering', particularly as the topics of the previous sentence do seem to belong to mathematics and philosophy departments. The prestige of theoretical science is as great as it was in Aristotle's time. Perhaps, and this will be part of my eventual defense of the claims of cognitive science, Aristotle's distinction between theoretical and practical/productive disciplines *has to be inappropriate* for the study of cognition.)

Nowhere is the paradox of cognitive science more apparent than in logic, particularly in the case of George Boole's *The Laws of Thought* (1854), arguably the first real advance in a subject that Immanuel Kant, in the 1790s, authoritatively proclaimed had been entirely completed by Aristotle. (Aristotle had indeed put together the first systematic large-scale study of it, though substantial parts of it had been worked up by previous thinkers and the Sophists made it, and tricky counterfeits of it, their stock in trade.)

Kant *did* have reasons for holding that Aristotle had successfully completed the science of logic, aside from the simple point that the medieval, pedagogically oriented version of it was still what was taught as logic in his own time. Kant could argue that Aristotle's system of classifying declarative sentences, and the immediate and syllogistic inferences they gave rise to, was systematically exhaustive. Moreover, if logic is to explicate the core of our reasoning powers, it would seem plausible that the first individual who made a concerted effort systematically to formalize this core might very well succeed.

Kant agreed with Plato and Descartes that geometry, arithmetic, and logic are native to the mind but he had a more convincing explanation as to why they turn out to be true of the world we experience through our senses. Kant claimed we knew about the world through two fundamental mental faculties.

Our logico-linguistic faculty affords us thoughts (statements, declarative sentences) and the deductive relations between them (the native portion of all this is the subject matter of logic). This faculty would tell us, for example, that

(1) IF IT IS TRUE THAT *it rained in Oxford at 2 P.M. on July 4th, 1862*, THEN IT IS FALSE THAT *it did not rain in Oxford at 2 P.M. on July 4th, 1862*.

Our senses, of course, would have depended on environmental input to decide whether or not *in fact* it rained then and there. (Lewis Carroll, on a now world-famous canoe trip with Alice Hart Liddell, later recalled a sunny afternoon, while the weather bureau reported rain.)

But there is an important *difference* in generality between the environmental inputs and the native structures that carry their information load. Environmental inputs are finite particulars and even a lifetime of them just mount up to a large finite number of them. On the other hand, the native faculty that generates (1) does not store this instance individually, along with an infinite number of others. It might well store a rule such as 'If a sentence is true, then its negation is false' or '$p \rightarrow --p$', or whatever our finite means afford us to grapple with the particular sentences the world deals us, arbitrary selections from an infinite pack. While it seems clear enough that we must have some kind of generalizing logical capacity, how it operates and how to characterize it still remain some of the deepest and most central questions of cognitive science.

Our other fundamental mental faculty, Kant maintained, natively structures sensory input to put together perceptions of objects in space and time. Recent experimental research supports much of Kant's view that we *do not* first feel input on our retinas, ear drums, nostrils, taste buds, etc., and *then somehow slowly learn to infer from this* a moving array of three-dimensional objects in space and time (this recent research seems to show that neonates have the notion of solid three-dimensional objects from birth). We sense the world around us as objects spread out into a spacial/temporal array, and while our eyes usually receive the most important environmental input, our other senses often make important, unnoticed contributions to this array. There are illusions, for example, that persist until one is allowed to touch, and then one 'sees' the array correctly, though input to the eyes has not changed. The blind use sound to construct a somewhat dim and colorless spatial array, but they usually are not conscious of the sound they use to construct their spacial impression. Descartes, long before, speculated that a blind man, thoroughly experienced in using his stick, might be said to 'see' with it. Kant, as at least a partial answer to Plato's problem, argued that geometry stemmed from our native, mental structuring of sensory inputs into a spatial array. Arithmetic, specifically the ordinal continuum of numbers, arose from the related native structuring of these inputs into a temporal array.

Recent work provides considerable tacit support for Kant's claim that these *two* faculties – the logico-linguistic and the spatial/temporal – are the most fundamental, native, human cognitive faculties. Leaving cognitive psychology and artificial intelligence aside, when you look at the specific studies that people think are making or will make crucial contributions to cognitive science, you find linguistics and visual perception.

George Boole, in reviving the study of logic, insisted, like Kant, that 'the laws of thought' (that is, the truths of logic) are separate from mathematics. The luxuriant structure of mathematics uses logic and must conform to it. But one can have logic without thereby having mathematics. Boole proposed to characterize the essential features of thought, the laws thought followed (looking forward, were a computer to follow these laws, then one might suppose that it was a thinker).

The most obvious paradoxical aspect of Boole's work is in his title. If there are 'laws of thought', then one would suppose that when humans think, they must conform to these laws, much as physical objects must conform to the laws of thermodynamics with respect to heating and cooling. Equally, if this were the case, one might think that Boole should have based his book on experimental studies of the actual intellectual behavior of humans, warts and all. But of course he did nothing of the sort. In fact, he acknowledged that humans often violated logic in the sense of saying or writing something that violated the laws of thought.

We can make sense of this in current terms through the views of the linguist Noam Chomsky and the psychologist Jean Piaget (both have disavowed the departmental labels I am giving them, so they might both be justly called, simply, cognitive scientists). Though in different ways, both claim that in normal maturation humans natively develop a logico-linguistic *competence*, of which a portion is modeled by logic. Through slips of the tongue, through problems of memory, ambiguity, or processing length, through the heavy pressure of desire or cultural prejudice, etc., deviations from this competence may well occur.

(Someone may say, 'Competence, shampetence!, you are just talking about cultural and class prescriptions.' There is an answer to this accusation. What differentiates this sort of competence from such is that logico-linguistic competence seems to arise in every human community, by much the same stages and with many of the same underlying features, mostly without any need of actual instruction, correction, or training, only exposure to the local physical environment and language, perhaps, being necessary. Further, individuals, like Meno's slave boy, who deviate superficially from their tacit knowledge, or unconscious competence, can to some degree be brought, by questions or examples, to see that they have deviated from a knowledge that they did not consciously realize they had. To the contrary, cultural prescriptions usually have to be drummed into people explicitly, vary enormously from culture to culture, and lack an infinite, generative character.)

Oddly enough, what has most distinctively survived to the present in Boole's logical work is what he fashioned that applied to the understanding of computers, not humans. Along with Babbage, Boole did much to overcome the exclusive British mathematical commitment (dating back to Newton's calculus notation as against Leibniz's binary one) to the familiar ten-base numerical system, as opposed to the two-base system now used by electronic computers. He also used the '1' and '0' binary notation to stand for true and false. This led him to explain the logic of operators like AND and OR in such a way that his sentence logic and notation turned out to be just what was needed to explain computer innards, which would be composed of trees of 'flipflops' (off–on switches).

Boole saw logic as 'the laws of thought' and distinguished it from mathematics. Gottlob Frege, who denied both claims, is credited with the two achievements that gave full birth to modern logic in *Concept-Writing* (1879). First, he devised the first fully adequate notation for sentence and predicate logic. Second, at the same time, he solved the problem of multiple quantification, which would give logic an enormous expressive power and generality that Aristotle's logic lacked. So enormous a power that Frege's ultimate, and credible, project was to show that the whole number system and all the truths of arithmetic can be expressed in purely logical terms, and indeed, all simply follow as theorems from his purely logical axioms.

'All men are moral' is a singly quantified sentence of just the sort that Aristotelian predicate logic triumphed in. 'Every boy loves some girl', which is doubly quantified, has no analysis in Aristotelian logic, whereas modern predicate logic, whether in Frege's or Bertrand Russell's notation, reveals more sharply than English the two possible readings (i.e., that there is some one single girl whom all the boys share a love for, or that for each boy there is at least one girl he loves, not necessarily, as you go from boy to boy, the same girl).

While one might not feel it important to uncover such erotic ambiguities, a large number of the generalizations that mathematicians, and other scientists, want to make require multiple quantification. These can be formalized with great clarity in modern logic. One major reason that logic languished, even after scientific research began to flourish again, was that the medieval Scholastics had set Aristotelian logic in churchly concrete. Hence, when modern thinkers found it inadequate, they did not (mostly) think to revise or expand logic, rather they ignored it. Perhaps, then, we had to wait until the secularized latter half of the nineteenth century before real, fundamental advances in logic could be introduced. But there was another reason for this change. Boole, to some degree, and certainly Frege and many other mathematicians, had by this time begun to worry about the foundations of mathematics.

In part these worries happened because of the great expansion of mathematics, and the more strange and less visualizable objects that came with this expansion (in the case of calculus, actual contradictions were found in traditional interpretations). The historian Spengler relates the anecdote that some ancient Greek sailors were disturbed by the presence of a passenger who talked about irrational numbers (such as the square root of 2 whose decimal expansion never terminates and so cannot be given through a ratio). The sailors, true geometricians, threw the passenger into the sea. How much more would they have reacted to talk of imaginery numbers such as the square root of −1, or bewitching talk to the effect that beyond the 'mere' infinity of the natural numbers, there is an infinitely more infinite infinity, and another infinitely more infinite infinity than that − and so on, *ad infinitum*, as Georg Cantor clearly seemed to prove. (This Cantorian 'heaven' of endlessly higher infinities so disturbed some mathematicians that they denied the straightforward existence of the first infinity of natural numbers, for they realized that if this were granted the rest had to follow. More of these intuitionists later.)

Aside from this luxuriance, the growth in mathematics made it possible to apply mathematics to itself, so that one might ask from outside a mathematical language, for example, whether that language could allow a contradiction, or afford a proof of every true sentence, and so on. Mathematics had become nervous and introspective.

In any case, Frege had an elegant solution. First, he presumed, as everyone, that logic could not have any internal contradictions. Then he proposed to show that one could deduce all of arithmetic from nothing but the axioms of logic (including the bare idea of a set or collection of things). *This* would show that arithmetic was free from contradiction. Frege spent 25 years on this deduction, first sketching and justifying the project in *The Foundations of Arithmetic* (1882). Leaving aside some refinements, the number 'one' could be explained as the set of all sets in which there was a member a, and if anything x were a member, then $x = a$ (this ensures that each set has only one member *without* using any numbers). 'Two' would be the set of all sets in which there were members a and b, and for any x, $x = a$, or $x = b$; and so on for the rest of the numbers. 'Zero', naturally, would be the set of all sets with no members. All this seems long and cumbersome, like balancing your cheque book with binary numbers. But the point, you see, is that Frege could define the natural numbers logically in terms of the properties of sets, without *presupposing* the numbers in the first place. Frege's project was supposed to culminate in *The Laws of Arithmetic* (1902). (Notice, parenthetically, that if number theory follows from logic, then our characterization of 'a thinking being' would seem to include knowledge of arithmetic.)

Frege, denying Boole and addressing a question we raised in chapter 1, claimed that logic describes the most general truths about the universe.

Thus, I suppose, Frege would have endorsed the astronomer's view that intelligent aliens (should there be any) could be expected to know the sequence of prime numbers. He above all wished to disparage 'psychologism', by which he meant the view that logical (and arithmetical) truths derive from our particular cognitive nature and are not true, or cannot be known to be true, apart from this nature. He blamed Kant, above all, for this view. Like Plato, Frege believed our minds 'saw' these eternal truths. Like Plato, he needed to explain how we could do this, how their truth could be guaranteed. Unlike Plato, Frege had the solution of showing that arithmetical truths reduced to logical truths; and logical truths were such that denial of them could be shown to be self-contradictory and senseless.

In 1902, Bertrand Russell, having read the galley proofs of Frege's intended *magnum opus*, sent a postcard to Frege, in which he pointed out that the unrestricted way in which Frege used sets to define numbers opened his system to internal contradiction. Russell had followed Frege's project with respect and interest, and was by far the major influence in making Frege's work known to the world – not only directly but eventually through carrying the project to one sort of masterful and sprawling completion in the publication, with A. N. Whitehead, of *Principia Mathematica* (published 1910–12, though Russell reports working on it since 1897). Frege reacted to Russell's postcard by at once admitting Russell was right. Though he added an appendix to his book outlining a technical solution for the problem, Frege soon came to feel that arithmetic after all needed grounding in the spatial/temporal mental structuring that constitutes vision. Russell did go on to produce a fairly natural but extremely complicated solution – the ramified theory of types – but most mathematicians who carried on with set theory preferred simpler but obviously arbitrary patches that avoided Russell's paradoxes.

Russell later gave a cheerfully concrete version of the evil beast that he first exposed in his postcard to Frege: *the barber who shaves all and only those in his village who do not shave themselves*. Suppose a rural 1910 English village. Suppose that all the villagers belong just to one of two sets: self-shavers and non-self-shavers. Now try to imagine a barber who fits Russell's specification. Is he a self-shaver? He cannot be in that set because he shaves only those who are non-self-shavers. So it follows that he must be a non-self-shaver. But since he is supposed to be the one who shaves all non-self-shavers, then he must shave himself and so has to be a self-shaver. But then . . .

Notice that we have a paradoxical noun phrase here, not a paradoxical sentence, as we had with Epimenides the Cretan. We have found a paradoxically impossible object as much as an unspeakable sentence. The general formulation Russell sent Frege comes into view if we replace *shaving* by *membering*. There obviously are many *non-self-membering* sets. The set consisting of all the three-letter words in this book is not a three-letter word; nor is the set of single-membered sets itself single-membered (there is more than one individual thing in the world). On the other hand, the set of all sets would at least seem to be a *self-membering set*. Hence, every set is either not a member of itself or it is a member of itself. But, now, what about *the set consisting of all non-self-membering sets*? If it is a member of itself, then it is not; but if it is not a member of itself, that is just what will make it a member of itself.

While Russell could thus put the barber paradox more generally, one still can feel that this species of pun, like the rest, belongs to the lowest form of humor. The threat to reason can be tittered away without Pauline curses. Russell himself, while he labored for years to find a proper way to exclude such destructive paradoxes from *Principia Mathematica*, was haunted by the feeling that the problem had to be trivial, that there had to be a simple solution, that such joke-like puzzles surely could not be the key and the impetus for a deeper understanding of the structure of logic, mathematics, and thought.

By the middle of the twentieth century, *Principia Mathematica* was the only twentieth-century book to be included in the University of Chicago's austere list of great books. Even this is insufficient to suggest the excitement that logic and Russell held in the first decades of the century for young men such as Norbert Wiener, Rudolph Carnap, Alfred Tarski, and Ludwig Wittgenstein. Russell and Wittgenstein, and the logical positivists, made compelling the view that logic gives us the 'deep structure' of not only mathematical but also scientific and human discourse in general. The underlying structure of thought appeared to have been found, and been found to be a language of great simplicity, unity, elegance, and power. Translation into this language, from the superficial and clumsy apparent babel of human jargons into radiant deep structure, would make clear what people really thought and said. Equally, the evaporation of apparent meaning would help clear off the emotive nonsense that bedevils our talk, particularly on ethical, religious, and philosophical topics. The mind seemed to be opening to view in the deductive march of symbols in *Principia Mathematica*.

(To remind one, in the midst of the sublime, of the more popular: over the same period of time in which Russell put together his masterwork, the populace wondered over the exploits of Sherlock Holmes, the first, and most enduring, scientist/hero of fiction, whom Arthur Conan Doyle presented as the master of deduction and logical analysis.)

It is time to stand back from this historical sketch to make four points before moving on to Alan Turing.

1. Russell's work made clear the centrality and power of logic for all cognitive study and, in doing so, suggested how it might both show us much about the content and the form of thought, in particular in comparison to human languages. Even more than Russell, the logical positivists popularized a sharp and exhaustive distinction between *analytic* and *synthetic* assertions. Analytic statements (or thoughts) are true by virtue of their form or meaning. Such analytic truths comprise mathematical and logical truths, and everyday logically true sentences such as 'All oculists are eye-doctors' or 'If this is a normal coin, your chances of tossing heads are one in two.' *Synthetic* statements (or thoughts) are true, if true, because of the data, because of the way things actually chance to turn out.

What this distinction suggests is that a good description of a mind (or a language) would be a specification of its analytic sentences, which would be in effect a specification of all the logical relationships among all the sentences in the mind's potential repertory. This would specify the *content* of the mind, all that it knows apart from the particular data of experience and memory. Since such analytic sentences are infinite in number they cannot be stored one by one. Rather the mind must have rules that generate sentences in general and rules that distinguish analytic sentences (or thoughts) in particular.

What rules? – hardly likely that they should include the same five (later four) axioms as those of the *Principia*, all in Russellian notation, but many have come to agree that they must surely include rules that bear some sort of specifiable relationship to the *Principia*'s rules, in particular that they generate at least roughly the same truths of sentential and predicate logic. Notice that we now have a moderate-sized and very abstract bit of a *competence* model of human thinking. Continuing the tradition, the model specifies what a normal human seems built to know, an underlying logical competence that can be brought to a more self-conscious pitch simply through questioning. Admittedly, extraneous performance factors, such as bias, interference, memory limits, etc., may sometimes mask a competence model. Further, the model may not specify anything about how the rules function in the actual production of individual thoughts.

None the less, the model strongly suggests that the production of thoughts, of chains of reasoning – *thinking* – must be like theorem-proving. The line of thought is well nigh irresistible. As data comes in (as our senses work), do these new data sentences, plus the old ones, establish something logically? – what deductive chain holds out something new, and is some possibility ruled out, etc.? To give an example, Sherlock Holmes stunned the inspector by mentioning the 'curious incident of the dog in the night-time.' 'But the dog did nothing in the night,' the inspector thoughtlessly exclaims. 'That was the curious incident,' remarks Sherlock Holmes, later explicitly arguing that since the dog would have barked if a stranger had taken the horse but the dog did *not* bark, therefore the familiar trainer must have taken the horse. Holmes' reasoning starts with some general factual assumptions about dogs and some particular data about a dog and the layout of a racing stable; from these he deduces his conclusion much as one would prove a (admittedly humble) theorem.[3]

2.   Perhaps most investigators are still convinced that something like command of predicate logic is basic to thinking and to understanding a human language. But there are problems posed by predicate logic itself and by the claims of other weaker and more luxuriant and more powerful logics. Moreover, our concern naturally shifts from what a thinker must know to *how it could have developed such knowledge* and to *how it constructs and employs such knowledge*.

You will recall the gap that Russell opened for Frege between predicate logic and the predicate logic + set theory that would produce the truths of arithmetic. If we suppose that all thinkers have predicate logic, do we also want to suppose that they have a command of arithmetic? Similarly, if we believe that command of a human language means a grasp of the logical relationships among its sentences, would this *also* have to mean a grasp of arithmetic (since you can put arithmetical equations into English or any other language)? Computatively, this might also be an important divide since we know that there *is* a proof for every truth of predicate logic, whereas this is provably not true of arithmetic.

But while we know that there is a proof of every truth of predicate logic, we have also established that *there is no mechanical procedure* that guarantees we shall find the proof in any reasonable number of steps (say, less than a trillion trillion trillion). Hence, whether speaking of humans or electronic computers, we know that a thinker will need some heuristics – some good though not infallible ways of looking for proofs. There are rapid guaranteed mechanical procedures for the simpler sentence logic that predicate logic incorporates. But sentence logic alone seems a much too limited model for a thinker. So if theorem-proving is suggestive of what thinking is, it must include heuristics. Our thinker must be a predicate logician *and* a good guesser. Others argue that still more limited, and hence less computationally demanding means, may do the real work.

Recently, logicians and linguists have emphasized the claims of modal – logics of possibility and necessity – and intensional or analytic ones as well. We distinguish statements such as 'It is green all over, so it cannot be red' from 'It is a swan, so it is white' (said in the European countryside or before the discovery of black swans in Africa). The first statement says what is necessarily true (or analytic), while the second rests on factual data. We humans seem to have a large stock of predicates with intensional relationships to each other; when we acquire English we seem to acquire a large number of analytic truths that arise from these relationships. Clearly, however, it is a daunting task to try to lay out these truths systematically, and it seems absurd to think that all thinkers must have the same stock of them.

We may even doubt whether humans, let alone other thinkers, share all of the classical logical truths of the *Principia*. I mentioned that some intuitionist mathematicians have questioned whether we should accept reasoning that depends on presuming the existence of actual numerical infinities (as opposed to 'infinitely proceeding sequences'). If you press their qualms, they will suggest that we should not accept a classical axiom of logic, namely, bivalence. Bivalence claims that we must assume that 'every statement is true or it is false' ($p \vee -p$). The intuitionists do not accept bivalence as proved for any statement unless there is either a proof of the statement or a proof of its denial. Similarly, the intuitionist insists that a proof that *there will be no disproof* of a statement *does not prove* the statement ($--p$ does not establish p). The intuitionist mind has a kind of caution, an insistence on a showing-tis-so, that the classical mind lacks. *You have to show me how you can build it*, says the intuitionist.

While some intuitionist logicians are just concerned with mathematical thought about infinity, others have claimed that human reasoning in general is intuitionist, imbued with an emphasis on constructive procedures. Most famous for his claim that children go through cognitive stages, developing from concrete and limited reasoning to more formal thought, Jean Piaget claimed that mature human reasoning is really intuitionist, rather than classical, in its logical character.

While Piaget's many studies suggest the central importance of logic in characterizing human thinking, one exemplary misunderstanding of it reveals how sorely fragmented the study of cognition is. Shortly after Piaget's death in 1980, a Harvard University psychologist, widely regarded as a leading Piagetian, wrote a summary of Piaget's work. He ended with a caution that perhaps Piaget might have been too committed to a 'Western' notion of thought and blind to 'Indian logic'. But the Harvard psychologist's description of the 'Indian logic' makes it clear that he is, unwittingly, referring to the two intuitionist departures from classical logic that I have just described. The 'non-Western logic' that Piaget is supposedly blind to is in fact the logic that Piaget himself espoused, particularly in a book he co-authored with the major Dutch logician E. W. Beth. While no one can be faulted for failing to read a particular book among Piaget's many, constructivist and intuitionist concerns are central to Piaget's work on cognitive development. Only someone whose disciplinary blinders had made him oblivious of logic could have failed to see that the mysterious East was the Netherlands and Piaget himself. We need cognitive science!

Indeed, perhaps both intuitionist and classical logic might turn out to be more like 'higher-level programming languages' that our brains can simulate only haltingly and roughly, with severe limitations, while our 'machine language' supports this simulation among others in a quick but much narrower and more specific grab-bag of tricks?

3. As many others, Russell often asserted that the surface grammatical form of sentences could mislead one about their underlying logical form or deep structure. Translating natural language sentences into the notation of predicate logic would dispel paradox and misunderstanding, along with specious metaphysics. But one can ask *which notation*?

In previous paragraphs I have written of the truths of logic and mathematics, and of thinking as theorem-proving, so as to *ignore* anything about the linguistic or mental *representations* in which these truths are thought or said. This stance is in the same spirit as that of the astronomer who supposes that intelligent aliens will also know essentially the same logical and mathematical truths as we – though the physical and psychological materials and structures, and the accents and conventions of their languages and symbolisms, may be vastly different in a number of dimensions from our own. The alien is expected to notice regularities in the physical patterns in the signal we send (whether in radio waves or incisions on flat metal or whatever). What the astronomer hopes is that, through some combination of deduction and heuristics, aliens will be able to 'crack the code' by extracting a system of representations that can be, at least roughly, translated into the alien's own language. In its initial stages, the message's representations will exhibit basic logical and mathematical arrays and formulae in such a lean and salient fashion that the aliens will infer both what the arrays and formulae mean and how the representational system works. This much gained, the message will continue to piggyback on itself, depending on presumably shared physical knowledge, to give the aliens some description of ourselves and our local world.

Let me give a simple concrete example. The astronomer's message might begin by giving the first 20 prime numbers through radio bips:

one bip, pause, two bips, pause, three bips, pause, five bips, pause, seven bips, pause, eleven bips, etc. Then this prime sequence might be repeated but in a binary notation, employing a bip for 1 and the longer bop for 0. By exhibiting arithmetical and logical truths, one could communicate representations of numerical, geometric, and logical notions; one could also communicate a grid with coordinates, followed by two- and three-dimensional pictures, etc. Our bip/bop notation is most austere, the one sign differing from the other in only the length of electromagnetic activity (a difference like that of the off–on flipflops that are the ultimate constituents of our electronic computers). If we move to the visual (graphically, 1, 0) notation of binary numbers, we increase the complexity of the signal; if we move to the visual, ten-base arabic numbers, we should pose aliens considerably greater problems of intepretation, and a shift from arabic notation to English numerical expressions would create further burdens. Notation matters. And it is inescapable, for whatever Plato supposed, there is no cosmic deep notation in which all formal truths are recorded for all time for the employment of all species of thinkers.

The importance of notation is curiously evident in Russell's own case. He owned a copy of Frege's *Concept-Writing* from his undergraduate days but only, as Russell put it, after years of related work in other notations, was he able to read and make sense of Frege's arcane and since unused notation. Russell reacted differently to another notation at the International Congress of Philosophy in Paris in July, 1900, writing in his autobiography

It became quite clear to me that his notation afforded an instrument of logical analysis such as I had been seeking for years, and that by studying him I was acquiring a new and powerful technique for work that I had long wanted to do. By August I had mastered it . . . I spent September in extending his methods to the logic of relations. It seemed to me that every day was sunny . . . For years I had been endeavouring to analyse the fundamental notions of mathematics, such as ordinal and cardinal numbers. Suddenly, in the space of a few weeks, I discovered what appeared to be definitive answers to the problems that had baffled me for years. I was introduced to a new mathematical technique, by which regions formerly abandoned to the vagueness of philosophers were conquered for the precision of exact formulae. Intellectually, the month of September, 1900, was the highest point of my life.[4]

We are again reminded that the invention of cognitive technology, from the Greeks' alphabet and the Arabian numerals to Russell's notation and LISP, can seem a major theoretical advance, a carving of cognitive nature at her proverbial joints.

4.   As a match for the near endless possibilities of notation, there is the

unavoidability of adopting some narrow set of representations, for reasons rooted in one's sensory organs, cognitive wiring, and cultural and historical accident. When Russell supposed that the surface form of English sentences could mislead one about their logical deep structures (as these might be represented in Russell's notation for predicate logic), *how intimate might the relationship be between English language sentences and predicate logic formulae in Russellian notation*? Is it merely like that between some intelligent Alpha-Centaurian octopoids' communication mode structures (gigablips, xyzles, or whatever) and predicate logic formulae *in Russellian notation*? Or is there a much more intimate connection, with the formulae closely mapping deep structural relations between human linguistic representations? (Polish logic notation, like computer LISP and unlike Russellian, puts logical connectives before what they connect, so we have &vpqr, rather than (p&(qvr)); at least initially, most humans find Polish very difficult to work with.)

The ancient injunction, *know thyself*, still sets us a formidable if exciting task. The last 100 years look to have provided us more progress than the previous 2,000. More importantly, we seem for the first time to have the wherewithall to do the job. As Plato recognized there is for us no enterprise more fitting or more frightening.

Here, truly, is an invitation adequate to our capacity for wonder, for research and creation.

4

Meaning Must Have a Stop

In consequence of A. M. Turing's work, a precise and unquestionably adequate definition of the concept of formal system can now be given, the existence of undecidable arithmetical propositions can now be proved rigorously for every consistent formal system containing number theory. Turing's work gives an analysis of the concept of 'mechanical procedure' (alias 'algorithm' or 'computational procedure'). This concept is shown to be equivalent with that of a 'Turing machine'.

Kurt Gödel[1]

Alan Turing (1913–54) commands our attention as the first complete exemplary cognitive scientist, the one who made the requisite foundational moves in several disciplines and made them seem to form a single cognitive science.

He was the first in that in 1937 he published the most foundational characterization we have of computation and its possible forms. Hence, cognitive scientists speak of Babbage's unfinished Engine, of electronic digital computers, and of ourselves, as *Universal Turing Machines*. Many suggest that when we want to say that organisms or machines of very different physical composition *have the same particular cognitive state* (belief, ability, etc.) what we are really saying is that they both realize the same particular Turing Machine.

He was first too in that he had a major part in the development of one of the first electronic digital computers, *Colossus* (1944), and had a hand in both of Britain's postwar computer construction efforts. He also, in various ways, contributed to the earliest efforts at programming and development of higher level programming language. He was first in emphasizing the most basic claim of cognitive science: that thinking is thinking whether realized in human neurology or microchips or whatever. Similarly, we now speak of what is widely regarded as the fundamental standard for determining that an actual computer genuinely thinks as the *Turing test*.

How might we classify Turing's achievements in Aristotelian terms? Is Turing a discoverer or an inventor; a theoretical mathematician, an engineer, or, recalling Oscar Wilde's epigram, an artist?

As Aristotle noted, we think theory more profound than practice, science more basic than engineering. Yet sometimes the only adequate way to show that something can be done is by actually doing it, the only adequate way to show that something can exist is through actually constructing it.

To sketch the growth of a science, it helps to fix on a few individual contributions, whether in the form of theories, constructions, demonstrations, arguments, or experimental results. This inviting simplification is perilous, however necessary, particularly when research has proceeded in parallel at many centers, particularly when priority, influence, and the direction and value of further work are under debate. To put this most negatively, some have foolishly regarded Turing's 1937 result as revealing decisively the limitations of machinery as opposed to humans. Similarly and more justly, the German engineer, Duze, who failed to meet Hitler's standards with his funding request in 1942, none the less ended the war providing the V2 effort at Pennimunde with something like electronic computation. And the American ENIAC effort of 1944–5, though touted patriotically in the US as the only contender, has had its priority vitiated by the 1973 court decision that provides a now widely accepted case that Iowa State Professor John V. Atanasoff quietly produced the first electronic computer in 1939 or so, only to have his work transformed into ENIAC without acknowledgement by John W. Mauchly.

In face of such contention, we might agree with Alan Turing that Babbage invented the programmable digital computer and that, as Turing ended his 1950 paper, 'We can only see a short distance ahead, but we can see plenty there that needs to be done.'

Finally, there is Turing's specific paper describing what we today call the 'Turing test' for whether a computer *thinks*. Certainly he makes a central and deep philosophical point (a rediscovery and skillful expansion of Descartes' proposal). But he presents it in a flip and brash manner, as if the point of cognitive science is to give human chauvinist pigs a belly-ache. A sentimental transmigrationalist might suppose that La Mettrie had reappeared. (This 1950 paper also lays out a startlingly prophetic prospectus for cognitive science, much of it still fresh and relevant to today – and tomorrow.)

Personally, Alan Turing was a cheerful, unassuming man, a Cambridge University pure mathematician who spent much of his time physically constructing a prodigal variety of gadgets, from the most *ad hoc* and gimcrack conveniences and toys to wartime code breaking and communication technology. He became a perpetual undergraduate, defiantly and naively honest and hopelessly inept socially, who might show up for work for weeks with his trousers held up by string, happy that he had found the least time-consuming solution to the missing belt problem, or who might disturb the rector's dinner party by innocently insisting that it is obvious *heresy* to say that computering machines cannot really think because they cannot have souls *because this would be a heretical denial of God's power to ensoul whatever He chooses*. Others coined the terms 'Turing Machine', 'Universal Turing Machine', and 'Turing test'. Others, including the Library of Congress, have given him the backhanded compliment of *lower-casing* him, writing 'turing machine', as if he were not an individual human being but rather an idea or a kind of artefact, what biologist Richard Dawkins has called a 'meme', the cultural equivalent of one of our self-replicating DNA, one of – to further quote Dawkins – our 'selfish genes'.[2] Exemplifying our paradoxical frontispiece, one Turing meme, the Turing Machine, provides the foundational characterization of the nature of memes and genes: we learn that nature can picture itself too.

Let us look at Turing's memes.

Turing Machines

Therefore, paradigm-testing occurs only after persistent failure to solve a noteworthy puzzle has given rise to crisis. And even then it occurs only after the sense of crisis has evoked an alternative candidate for paradigm.

Thomas Kuhn[3]

La Mettrie proclaimed that thought was a natural property of 'organized matter'. But he did not survey this supposed property in any detail: he did not purport to show how a machine could do the job. Alan Turing did in the paper he wrote as a Cambridge University student in his early twenties, 'On Computable Numbers, With An Application To The Entscheidungsproblem'.

At the time, as the Kurt Gödel quote that began this chapter suggests, this *application to decidability* held center stage. Turing rigorously proved that there had to be undecidable arithmetical propositions, propositions neither provable nor disprovable in any formal language or by any other acceptable mode of computation. Further, there could be no reliable, systematic way to identify and quarantine such propositions. (As Turing recognized, this also meant that no feat of formal language design could protect a powerful computational device from unwittingly attempting a computation that would never halt until physical breakdown.)

But while, at the time, the undecidability result seemed the most startling discovery, his foundational achievement was even then tersely recognized by Gödel: to have characterized precisely the intertwined notions of *formal system* (or language, formally described) and *mechanical procedure* by showing them to be equivalent to the notion of a 'Turing Machine').

Gödel suggests what this transformation may mean by his parenthetical comment that *mechanical procedure* (or *machine*) is 'alias "algorithm" and "computational procedure"', thus hinting perhaps that these latter notions rest on and are rigorously explained by the former (quote La Mettrie, 'Thought is so little incompatible with organized matter that it seems to be one of its properties on a par with electricity, the faculty of motion, etc.'). Sentences have form and formal meaning only within the context of processors and interpretive mechanisms.

In 1931, Kurt Gödel himself shook the mathematical world by showing that Russell's *Principia Mathematica* (*PM*) had to contain true but unprovable propositions. This crumpled a central supposition of Russell's project and undermined the unquestioned ambition of every mathematician from Euclid to Frege: to put together a list of axioms by which you could prove in a clear, step by step way *every* truth in any or every branch of mathematics. Gödel's results also threatened Russell's deeper claim about what numbers *really are*, namely, useful abbreviations of logical truths.

Russell, coming out of the empiricist, antidualist and antimaterialist tradition, naturally felt that numbers *could not* be basic, independent, parts of the universe like foundational physical particles – *could not* be quirky 'foundational supernatural particles', so to speak, which would form the mentalist half of dualism, the pure structure of mind as against mere sensory experience. But if *PM* had to contain true but unprovable statements, something about numbers would seem to be uncaptured in purely logical and physical terms.

In the context of arguments that suggested that dualism was incoherent and that empiricism had been triumphantly confirmed by experience, science, and argument, *PM* was supposed to be the final, triumphant, detailed and comprehensive, formalized demonstration of the complete failure of the most serious argument against empiricism, namely, that numbers have to be metaphysically basic because nothing else explains them and nothing else can replace them. What irony if *PM* itself becomes the very formal construction that makes it possible to prove numbers cannot be dispensed with!

Gödel formed the foundation of his remarkable proof by strange-looping Frege's project, as realized in Russell's *PM*. Arithmetic, indeed, could be logicized, so that all of *PM*'s propositions employ only logical symbols (though, of course, they create an image of the natural numbers and their properties). But, Gödel showed, you could also *uniquely number* the symbols, propositions, proof sequences, or anything else expressable in Russell's *PM*. This numbering system is often called 'Gödel numbering' and the numbers, 'Gödel numbers'.

The logical properties of *PM* *must* now have expression as relations between these Gödel numbers. *Inevitably* there must be an image available *in the purely arithmetical relationships between selected natural numbers* that pictures the properties of the formalized system of *PM* – *including* those properties by which *PM*, in turn, is supposed to formalize and completely secure all the mathematical properties of the natural numbers.

By Gödel's simple but tedious mechanical procedure, a small set of large natural numbers can be seen to encode the logical symbols of *PM*. A very large set of still much larger numbers will number all of the formulae, composed of these symbols, that *PM*'s grammar allows. It will be perfectly definite which Gödel numbers actually number these sets. Similarly, the provable theorems of *PM* must constitute a definite set of Gödel numbers and the unprovable ones must do so as well.

However, since this Gödel number structure is just part of the natural numbers, and the natural numbers are what *PM*'s logical language characterizes, we must be able to translate our Gödel number characterizations such as *numbering formulae as structurally unprovable in PM* back into the language of *PM*. Hence there has got to be an expression in the purely logical language of *PM* that asserts '*n* is not provable' in which *n* must be the Gödel number of that very proposition, '*n* is not provable'. But on pain of contradiction, this proposition must be *true* and must *not be provable* in *PM*.

While, so to speak, *PM* can *mouth* a lengthy characterization of the structure of a formal system that contains a true but unprovable characterization of itself, *PM* cannot know (cannot prove) that it is talking about itself*. In short, *PM* (and perhaps other powerful cognitive systems as well) must contain a formula that asserts *I am unprovable*. The really nasty catch is that *PM*'s structure makes it absolutely clear that this unprovable formula *has to be true* and that it is a mathematical truth.

Since the Frege/Russell project was to show that all mathematical propositions could be fully expressed and the true ones proved in a *PM* formalization, Gödel's incompleteness results determine a postcard to Russell longer but even more decisively destructive than Russell's to Frege. The Gödel number of the unprovable but mathematically true proposition is, truly, the 'Number of the Beast'. If we add axioms or change other details in the formalization of *PM*, the 'Beast Number' will certainly change but the method for identifying the new one is effective enough.

Indeed, particularly with the hindsight of Turing's work, we now know that the only *consistent* way to enrich *PM* so that all arithmetical truths can be proven is to add an *infinite* number of axioms, which makes the enterprise circular and pointless. Another, devilishly simple 'enrichment' would add an axiom *inconsistent* with *PM*, which means we shall be able to prove a contradiction and, from that, prove all arithmetical truths. This, as other devilish contracts, is much worse than pointless, for our crazy system (our mindless mind) will now, also, prove all arithmetical falsehoods, and there will be no way of telling which is which.

A Pauline, more from exasperated boredom than righteousness, might well say,

That strikes me as the most elaborate shaggy pun I could ever hope not to have heard. Indeed, I just now realize that I actually have not heard it because Gödel's paper just gives some directions for constructing the pun *if* you had the many years needed to construct it and a like time to listen to it. With Epimenides' one-liner, you laugh and groan, and go your way, agreeing that such infectious nonsense has no place in rational discourse. I can even see why Russell took elaborate steps to exclude related, contradiction-breeding constructions from *PM*'s formal system, though I am sorely tempted to quote Emerson's line that,

A foolish consistency is the hobgoblin of little minds,
adored by little philosophers and divines.

But Gödel's argument *assumes* that *PM* is *already protected* against this: the beast proposition is only unprovable in *PM* because *PM* is consistent. Who is to worry that we cannot prove a formula in *PM*'s language that is longer than the whole written text of *PM*, one which fully expresses the properties of a natural number which you would need miles of paper to write down?

The second part of Gödel's theorem answers our questioner by showing that another proposition you cannot prove in *PM* is that *PM is consistent*. Showing again how a step by step procedure would eventually produce the Gödel number of a proposition in *PM* that says, in effect, 'I am consistent', Gödel shows that if this could be proved in *PM*, the 'I am not provable' proposition will be provable *and PM must be inconsistent*. Hypothetically, if *PM* could prove its own consistency, then it would be inconsistent and everything would be provable; *therefore PM* cannot prove its own consistency.

(Assuming that a formal language is like a mind and theorem-proving like thinking, Gödel showed for *PM*, and Turing showed more generally, that there must be *unthinkable* thoughts in that, although *PM* can form the symbols that say *I am unprovable* it cannot know what they mean; further, perhaps, a mind complicated enough to ask the question *Am I consistent?* cannot have a noncircular answer.)

Since the most basic point of *PM* was to ensure the consistency of mathematics by showing that its supposed truths could be expressed, and proven true, in purely logical language which did not presuppose anything, Gödel's result is enormously troubling. Perhaps there is some other formal system which will do the job? Is there some other way to characterize numbers? Does this result require a new way of understanding logic and mathematics, and so, perhaps, thought, language, and mind?

It is these questions that Turing's 1937 paper addresses.

Our frontispiece illustrates this enlightening transformation. We may regard the artificial hand as the formal system which Russell drew in order to reflect on it, and thus to clarify and ground the natural number system (and so by implication natural mind). Gödel shows that the natural hand (that the artificial one sketches) must, in turn, contain an accurate representation of the artificial hand: *this* shows that the artificial hand cannot draw an aspect of the natural hand, namely, that part that allows the natural hand to draw that aspect of the artificial hand.

There is of course a certain paradox undermined bias here. Since *PM* certainly can picture the natural numbers, one could see the creator/creation loop that way: the natural numbers cannot have the expressive power to draw an artificial hand sufficiently well designed to draw its drawer. Only natural prejudice casts artifice, not nature, as limited.

Though, causally speaking, humans made computers, just as many arithmetical languages proceeded *PM*, the fundamental relationship is correspondence, mutual intermirroring. Such 'relative consistency' proofs have been found to relate a variety of formal systems: A is consistent if B is, C if A is, and B if A, etc., daisy-chaining along like Alice and the Red King. But these incomplete results not only make these interdependences ultimately ungrounded, these interrelated images are also unstable and at war. Gödel's construction procedure supposedly demonstrates mathematically that the *PM* formula is true. But, surely, that amounts to a proof of some sort and, worse yet, what we seem to have proved is that this proof of some sort is quite suspect, almost, as it says of *PM* after all, evidence that there is something fishy about this proof of some sort, evidence that if we have managed it, there is something inconsistent about the natural numbers.

Surely, we must seek a more precise, neutral, and basic formulation in which we undercut arguments about which image reflects which, and which must fail to describe which.

Mathematicians packed nightmares about consistency into the bare material requirement that it be shown that the formal system will never literally produce the physical symbols '0 = 1', an emblem of inconsistency. If we can show, so the thought goes, that a completely mindless following of explicit rules – the sort of thing the dullest and most literal-minded human could do if he just kept at it – will produce certain physical symbol sequences and never produce others, this will produce some kind of reassurance. *If we could prove that these literal symbol transformations can never produce the symbol sequence 1 = 0, then we are all right. The thoughts, abstract objects, Platonic forms, or numbers can take care of themselves as long as these literal marks on paper will never be produced.* We must, so the thought goes, produce purely formal systems, in which syntactical structures are related to each other without concern for meaning. However, as Gödel showed for *PM* and Turing showed more generally, the goal cannot be attained in a formal language (or mind) that expresses (or understands) basic arithmetic.

Paradoxically, though we have built formal systems as reflections of our thinking so that we should have an explicit demonstration of our own consistency, what we learned, by reflecting one in the other, is something quite unexpected and astonishing. The construction of the other is not only necessary to self-knowledge but the knowledge it brings is sobering and inevitably *mutually dependent*: I can see in the language (or thought) of the other what the other cannot, and, equally, the other has this knowledge of me that I can mouth but cannot understand. Here is a formal analogy of our everyday realization that others often understand one better than oneself, that what one says, means, or thinks, may (even must) be clear only to another.

What then could be more basic, what could both resolve the general question of consistency and the shuttle-cocking of hands, each picturing the other?

What is common to both hands, to both *PM* and its Gödel numbered reflection as they shuttle paradoxically, is *drawing*. Why not mechanize the picturing operation itself, why not undercut the question of what numbers are or symbols mean by asking what can be *done* by simple machines, what can be drawn, copied, counted, sorted, etc.? (The Art through which one could, with Wilde, draw Nature.)

The formalist program demanded formal languages in which proofs would be given explicitly, each step tagged by a justifying rule, so that the most dull and literal-minded human could follow the individual steps, most literally perhaps, with no understanding of what the formulae meant, only understanding that particular symbols must be replaced by others, as line succeeded line, in accord with a very brief instruction book, enacting an algorithm or computational procedure. The formalist program had hoped for a proof that there could never be a sequence of physical symbols such that the last step produced symbols '0 = 1'. Why not, Turing effectively proposed, instead ask what a machine *could* do and since the question of what rules require (or what formulae mean) is indeterminate, *why not simply build them into the machine itself, which will simply DO what it does and not mean anything at all or be FOLLOW- ING any rule.* ('Explanations come to a stop' as the philosopher Witt- genstein would put it; 'there is a last house in the lane'.)

Instead of thinking of mechanical procedures as ways of checking formalized proofs which in turn check mathematics, let us think of mechanical procedures, which is to say machines, as the most basic subject matter of mathematics. In the beginning was *counting*, an action of a natural machine, and word and number are extracted from this. Numbers and *PM* sets of sets are really just kinds of machines. Un- provable propositions and uncountable numbers are a particular kind of necessarily unstable or unspecifiable machines.

Turing's project of making symbols, numbers, proofs, and procedures into machines proved astonishingly successful. Since 'algorithm', 'com- putable function', 'formalized procedure', and 'formal system' and the like (and language and mind too) had no precise definition, Turing could only describe a most simple kind of minimal machine, and then show that anything that had been labeled by terms like 'formalized procedure' could be done by one of his machines, amongst which there was also a universal machine, one that could copy and so become any of the others. Turing showed, as Gödel's assessment handsomely states, that the real foundation of Gödel's results is the undecidable, the constructing or drawing that cannot determinately terminate, the machine that cannot be built.

The 'Number of the Beast' is uncountable because the Beast is a kind of finite machine that has to be mechanically unstable (Turing put it this way, thinking of the real numbers; alternatively, the point can be put as that of identifying computations that can never halt). The first 'beast form' is the machine that (supposedly) can check whether any T-Machine is, or is not, satisfactorily unparadoxical and determinate. Each T- Machine has a number that codes its procedural structure, but the checker machine, if asked whether its own number numbers a satisfactory machine, cannot give an answer, cannot have a determinate destination. Further, we cannot make a machine that can determine whether any particular machine (or equivalent computation) will ever print a given symbol (0, for example) – and this but introduces several undecidability results, several paradoxical machines. What Gödel had really discovered is that rote procedures that are the origins of numbers are unavoidably their termini. Numbers, whose mysterious power made the ancients think them gods, are not, foundationally speaking, eternal forms, abstract ideas, physical symbols, or sets of sets. Numbers are kinds of mechanisms, machines that *are* various mechanical procedures.

Having made mechanical procedure not a checker or illustration or byproduct of mathematics but its basic subject matter, Turing had to ask himself an extraordinarily new question. How can we determine from the ground up what discrete machines, machines that enact mechanical procedures, can do? How can we thus make the notions of algorithm, effective procedure, etc., precise?

In the intuitive account we start with input symbol sequences, then we manipulate them by the rules in our rule book until the output symbol sequence is produced. Since the output is material, we shall need a *printer*; for similar reasons we shall also need a *reader*, something that discriminates marks. Turing was well aware that the *reader* could be thought of as taking the sensory input to a thinking thing, as our brain must somehow read the neurological impulses from our sense organs; while the *printer* would be the motor output, like the neurological emissions of our brain that direct our musculature. Alternatively and more specifically, reading could be *hearing* and printing, *speaking*. And *thinking* is what goes on in between.

How complex does the input need to be, or alternatively, how many discriminations does the *read head* have to be able to make? Turing showed that multidimensional, or multichannel, inputs always can be reduced to much, much longer one-dimensional input, just as a live TV broadcasting system converts the image on the camera's lens into a single serial sequence of extremely rapid, digitalized pulses. It follows, of course, that just as the broadcast signal, long but very rapid like a tape which inputs one symbol at a time, can be disassembled into the picture, the two-dimensional pattern of pixils on your TV screen, so Turing also showed that any particular two-dimensional output can in principle always be specified by a single serial stream of signals. More generally, the point is this. We, obviously, often seem to get a lot of simultaneous multidimensional and multichannel input to our cognitive apparatus, but since this informational input provably can be transformed into a single sequence of signs, a machine with only a single input channel could get all the information that our brains get. Turning the other way round, there must be a single serial stream that will carry all the information needed to direct our musculature. All the motor output instructions that produce voice, expression, gesture, fingering, gripping, and the rest always *can* be given in an admittedly enormously long single sequence of symbols.

How many distinct symbols do we need on the input tape? How many distinctions are the minimum needed for any sort of computation, for any sort of information processing? Though his paper dwells on alphabet input machines, Turing also showed that two distinguishable marks (1, 0, for example) and blank were sufficient. (We humans, interestingly, seem dedicated to roughly 25 input/output natural language symbols, though each human language varies a little in the phonemic expression for these distinctions. Particularly given evidence that human sign language, surprisingly, naturally strives for something like the same number of signal contrasts, gesture phonemes so to speak, there is a lot of attraction to the thesis that our *cognitive* processing calls for this digitalization and restriction on symbol number. It is *not* as if the mechanics of our voice box and ear demand this step. If our mind/brain could operate *efficiently* with the input, we could speak a two-sound language like the binary 1, 0 of a digital computer's machine language; on the other hand, our voice and ear seem capable of producing and hearing hundreds of distinct sounds, so we might have languages with several hundred 'phonemes' if our brains could handle them.)

Turing envisioned then, in relentless minimality, a tape under reader and printer on which not only input and, eventually, output appear but also scratch-paper space for intervening steps. The machine can erase or print on the space under the head and move forward or backward by a square and which of these it does is completely determined by the single symbol or blank under its reader and by which of a small number of internal settings the machine is in. Each step the machine takes, then, can be given by specifying the read symbol, the internal state, the action of printing, erasing, or moving one square forward or backward, and the new internal state.

Another even more fundamental source of minimality is that the input tape can, in effect, specify not only data but also the instructions for transforming the data. Machines can be chained together to form larger machines and these in turn, and so on (to speed calculation by adding internal complexity, electronic computers build a lot of basic operations into the central processor; many call these *Von Neumann Machines* since Neumann designed the first well-known US computer, but the tradeoff technique is evident in Turing's paper and in his own work on what he called 'practical computing machines'). Further, the Universal Machine, which can be chained into larger machines, does this trick more globally:

it is the minimal machine of this format that, given an input description of a specific Machine, turns itself into that one. Using these clever hacks, explicit mechanizations of the construction procedures of *PM* and effective Gödel numberings, Turing demonstrated that some Turing Machine enacted everything that has ever, intuitively, been thought computable or mechanical.

'Gimcrack' seems the word for these last two hacks of Turing's, toys built into toys built into toys, etc., in which technological utility and practicality are maximally surrendered for minimality, in that the simplest arithmetical operations will consist of thousands of steps. But by describing the absolutely minimal Universal Machine and showing how everything else can be farmed out in a standardized sequence of *1*s, *0*s, and blanks, which is now called a Turing Machine Table, Turing gives us a fundamental, standardized way of describing any computation or computational device. Indeed, we now may say that if you, I, and an intelligent extraterrestrial octopoid *are in the same mental state*, what that means is that we may each be characterized by the same particular *Turing Machine Table*. Of course, since a literal, minimal Turing Machine makes thousands of steps in performing simple arithmetical operations, we essentially should regard Turing Machine Tables as the standardized specifications, or specs, for operations or machines which in practice must be achieved through a much more complex (and hence speedier) internal design, which will build in much of the tape.

(It is startling and ironic that our tiny minimal gimcrack toy of computation, bred for purely mathematical motives, should have also been an invention that provided a template in which Turing could formulate, mechanize, design, and construct by stages electronic digital computation sufficiently reliable and so fast that it broke coded radio messages between U-boats and Germany in hours rather than months, eventually even capping the crescendo of U-boat successes, which would have meant the collapse of the British war effort within months, with a precipitous and utterly decisive increase in U-boat losses. The Enigma Machine's coded messages had the spartan minimality of short-wave radio dot/dash/silence (like a Turing Machine tape) and the message always began with a description of a machine (of the Enigma Machine's internal rotor settings that would be used to encode the message, and which, if the decoding machine turned itself to those same settings, would make it the decoder machine for the message that followed).)

If any formal system can be explicitly mechanized as a Turing Machine, so can any actual machine, nervous system, natural language, or mind *in so far as* these are determinate structures. Number objects and formal languages (and their interpreters and interpretations) only reach stable specification within fully mechanized procedures, within the bedrock of Turing's Machines (symbol crunchers, information processors). This suggests that the rest of our world and natural languages (and their interpreters and interpretations) require understanding of analogous sorts. The 'meaning' of words is found in their contribution to sentences, sentences in turn make sense within languages, and these in their role within the mechanisms of individual brains and perceptual/motor peripherals, and these in turn in differing respects both explain, and are explained by, biological, social, and technological mechanisms – and a natural world, within which these brains and their bodies function adaptively. Thoughts and perceptions *are not* sentences or pictures eddying in our heads like snowflakes under glass in a paperweight: rather, they are what they are in terms of their functional (ultimately mechanical) role in our computative mechanisms. 'It is only', wrote Wittgenstein, 'if someone *can do*, has learnt, is master of, such-and-such, that it makes sense to say he has had *this* experience . . . We talk, we utter words, and only *later* get a picture of their life.'

Before going on with Turing's project, I want to see if we can put the formalized results we have reviewed into a more human frame. What *should* these results suggest about everyday sentences and thoughts, about natural languages and human minds? We shall see, in chapter 9, that scientific linguistics arose and advanced through application to natural language of formal concepts, particularly decidability. Scientific linguistics started with the discovery that linguistic utterances *cannot be adequately described* simply as sound bursts, as nothing more than physical vibrations, but only as multilevel formal structures whose character is determined by role played in human cognitive processing. Computer science, naturally, also takes off from this rich theoretical apparatus. An electronic digital computer *is* a Universal Turing Machine, and whether a Turing Machine (alias 'computational procedure') will ever halt is obviously a real engineering problem.

But what of the forced switch from numbers and meanings to formalized languages to mechanical procedures or machines? Can we find some analog of this in more everyday terms? What is the decisive culmination of the search for mathematical foundations in the undecidability results, and hence in mechanism as mathematical bedrock, supposed to mean outside of mathematical logic?

There is something paradoxical in this question, and in this chapter, in this book, of course. The 'earth-shaking, foundational and seminal' Gödel and Turing papers are only available (together) in *The Undecidable* (1967), a book in which they appear as photocopy reproductions of their original journal publications to save money, given the small number of expected sales. Even the many mostly typographical corrections Turing made to the original *London Mathematical Society Proceedings* are also copied as a separate item rather than corrected in the original.

Yet reference to these incompleteness and decidability results and to Turing Machines and Universal Turing Machines pepper and often centrally motivate the literatures of cognitive science from cognitive psychology to computer science, from logic, philosophy, and linguistics to art and literature. Douglas Hofstadter's *Gödel, Escher, Bach: The Eternal Golden Braid*, a luminous, minds-on, circus of paradoxes and a brilliant introduction to computability – Pulitzer Prize winner and best-seller – has sold several hundred thousand copies, though it is claimed that the book has functioned as a fashionable cultural talisman for the coffee table, only the Escher prints surveyed. Similarly, Thomas Pynchon's bestselling novel, *Gravity's Rainbow*, produced an extraordinary and frantic but brief demand in bookstores for anything about incompleteness and undecidability. Since *Gravity's Rainbow* realized the arc of Pynchon's increasing critical acclaim as densely self-involuted High Art within the arc of fashionable popular recognition, *Gödel, Escher, Bach* has had company on the coffee table.

This reception is paradoxically appropriate, for Russell's rarely read but much discussed classic *PM*, and even more, Gödel's and Turing's inversion and generalizations of it, are literally *unwritten*. The literal printed volumes of *PM* begin by introducing the logical symbols and system rules, show how varieties of much more compact, and in turn compacted, mathematical talk can be constructured as abbreviations and abbreviations of abbreviations of *PM*'s real language. So, most of the literal text of *PM* is not literally written in the language of *PM* (good thing, too, as the unabbreviated version would be library long; as it was, Cambridge University Press had to ask Russell to cover some of their expected losses for publishing the 'abbreviated' four-volume version). Gödel's godel-numbered characterizations of *PM*, were they physically printed without abbreviations, far outgo *PM* in length, and since these are *of course* characterizations of the unabbreviated *PM*, they are exponentially longer still. Turing's Machines are more prodigal still, for in them an absolute and enlightening minimum of basic parts and functions is obtained, inevitably, by the enormous number of steps taken in characterizing the abbreviatedly symbolized stuff of logic and mathematics, and discrete, mechanical processes more generally.

There is some symmetry, then, in this series of three printed texts in which unbuilt machines, characterized by mostly unwritten tables, which ground and describe in stupefying length a literally unprinted, immense structure of numbers, which in turn represent the largely unprinted, yet four-volumed, *PM*. And some truth in the symmetry, too. The physically printed *PM*, through its systems of abbreviations, threaded together by proofs and sketches of proofs, is designed to satisfy a human logician's sense of what needs to be made explicit. Further, the printed *PM* was introduced and threaded through by English sentences that further specify, generalize, and draw sense from the pure, unabbreviated *PM* (which is unwritably long and unreadably dense for the human reader). Levels of representation, interwoven by mechanical procedures and glossed in natural language, form both medium and message, and make it clear that *PM* is a popularization from the beginning. Gödel's paper, even more, begins with a powerful informal explanation in natural language of the more technical sketch of how the Gödel number-picture of *PM* could be constructed.

Turing sees that the interacting levels, intertranslatable languages, reach a stable foundation only in machines that simply do what they do, building the interpretation into formal structures, so that nothing is left to the unspecified whims of human interpretation or to the interpretive instability Gödel demonstrated. But he, therefore, must have the task of conveying what the construction of such machines, and their shortcut, more complex machine equivalents, might mean in human language. An adequate specification of what Turing's 1937 paper means (or strictly *is*) would ultimately require that the relevant interpretive aspects of human language and human cognitive processing be built into the Turing Machine specification, and that specification, by some Turing Machine with a monotonous tape of astronomical length, is obviously much less intelligible to us than Turing's printed paper in English.

In formalization and mechanization we merely meant to make explicit natural language, perception, and thought – starting, of course, with the simplest and most tractable parts. Can we then make vivid a sense of what these results might mean, in a richly expressive natural language such as English, when these results are seen in the structure of everyday experiences of the relationship between language, mind, and behavior?

5

Dark Glass and Shattered Mirrors

Our investigation is therefore a grammatical one. Such an investigation sheds light on our problem by clearing misunderstandings away. Misunderstandings concerning the use of words, caused, among things, by certain analogies between the forms of expression in different regions of the language. . . . Some of them can be removed by substituting one form of expression for another; this may be called 'analysis' of our forms of expression, for the process is sometimes like taking a thing apart. But now it may come to look as if there were something like a final analysis of our forms of language, and so a single *completed resolved form of every expression. . . .*

*(*Tractatus *4.5): 'The general form of the proposition is: This is how things are.' . . . That is the kind of proposition that one repeats to oneself countless times. One thinks that one is tracing the outline of a thing's nature over and over again, and one is merely tracing round the frame through which we look at it.*

A picture held us captive. *And we could not get outside it, for it lay in our language and language seems to repeat it to us inexorably. When philosophers use a word – 'knowledge', 'being', 'object', 'I', 'proposition', 'name' – and try to grasp the* essence *of the thing, one must always ask oneself: is the word ever actually used in this way in the language game which is its original home? . . . What* we do *is to bring words back from their metaphysical to their everyday use.*

Wittgenstein

The Cognitive Naturalist

Let us call on Ludwig Wittgenstein, whose philosophy of mathematics class Turing attended in spring 1939, some two years after he had published 'On Computable Numbers'.[1]

Wittgenstein's first book, *Tractatus Logico-Philosophicus* (1919), introduced by Bertrand Russell, gave starkly brief and compelling expression to the view that all one can say (and hence think) are particular atomic propositions that picture the microstates of the world, though normally one *says these* in the summaries, or abbreviations of them, that form empirical science. Logic and mathematics show us the most general features of these summaries, expressing 'the general form of the proposition' (Wittgenstein's technical contribution to logic is the truth table mechanical decision procedure for the spartan sentence logic he thought exemplified his claims). *Value*, whether aesthetic or ethical, is *unspeakable*, having no expression in the literal facts that form the totality of the world.

While calling beauty and goodness literally inexpressible, Wittgenstein pays respect to the unspeakable in sentences of poetic economy and biblical resonance, later remarking that the *Tractatus* attempts, paradoxically, to show us the unspeakable through saying what it is not. The *Tractatus* even bans itself, for it ends by asking the reader to appreciate that the *Tractatus* is like a ladder that one climbs up to get a clear view, only to realize from that clear perspective that the ladder, strictly, was not there.

Wittgenstein had scathingly condemned as a 'shilling shocker' Russell's own attempt to explain philosophy to a large audience in *Problems of Philosophy* (1912). Indeed, Wittgenstein begins *Tractatus* by cautioning that his book is only for a tiny number of people, like Frege and Russell, who have already wrestled with its logical issues. Further, the only interested publisher agreed to publish such an obscure and unsaleable book only after Wittgenstein's promise to obtain an introduction from the world famous Bertrand Russell.

Paradoxically, the *Tractatus* is now recognized, and widely studied and read, globally translated and popularly issued, as a cardinal classic of twentieth-century thought and literature, several of its printed sentences spawning subfields of interpretive publication. The book, indeed, is read and quoted, even chanted, by a still wider audience, with little grasp of the logico-mathematical issues that motivate, and help make sense, of much of it. What Wittgenstein numbered as its seven cardinal sentences have been made the libretto of a modernist choral. They also appear on T-shirts, in artistic collages, and in novels – in Thomas Pynchon's apocalyptic *V* a character climactically deciphers *Tractatus* lines in mysterious radio wave emissions that shower from Earth's upper atmosphere, while in Iris Murdoch's sophisticated narratives, characters talk as though Wittgenstein's words were something one had to come to terms with, a profound station in the twentieth-century intellectual's rite of passage. Like the lines and diamond paragraphs of Jorge Luis Borges, who exhibits deep philosophical points in his literary pieces, Wittgenstein translates well into other natural languages: the poetry is in deep structure, not captive to the surface music of a particular language.

It is, however, the actual, printed, natural language sentences of the

Tractatus that meme their way through our culture, while *PM*, and Gödel and Turing's papers, as loose sketches of effectively imagined sentences, languages, procedures, and machines, call out for further rendering and explanation, not textual exegesis. Since it is the ideas and constructions more than the actual words that meme outward, we can expect further renderings into natural language that are more than 'mere' popularizations.

In the lecture notes for the class Turing attended and in other writings, notably his *Philosophical Investigations*, Wittgenstein distinctively exhibits a concern with completeness and undecidability results, particularly as they confirmed, suggested, and reflected, his more general worries about mathematics and formal languages particularly as they in turn extend to natural languages and human minds, thus forming the central theme of Wittgenstein's later philosophy.

Interpretation of Wittgenstein's posthumously published and avowedly unsystematic later work is a thriving, largely hermetic cottage industry, and I do not wish to add to it. I shall quote his words, and arguments commonly attributed to him, to show you a broad justification, through exposure of the anomalous structures of everyday language and mind, of what Turing *did* in his 1937 paper and thereafter, in theoretical and actual machine building. By the same token, Wittgenstein's later work also provides a vivid revelation of the problems that cognitive science must face and the fallacies it must resist. Wittgenstein himself intended only to expose the anomalies of everyday cognitive life, and the incoherences of the traditional view of mind as consciousness that rises up from it. But it is precisely those incoherences that the new cognitive science casts out and those anomalies that it so revealingly explains.

There seems no clear evidence that Wittgenstein ever read Turing's paper.

A masterful, intense figure, who lectured without notes while a select few professors transcribed his words, he allowed little in the way of independent questions or counterargument. 'He was perhaps', Russell wrote over 20 years after he first got to know Wittgenstein, 'the most perfect example I have ever known of genius as traditionally conceived, passionate, profound, intense, and dominating'. After four years as Russell's student and most intense interlocutor, young Wittgenstein passionately savaged Russell's latest and most ambitious work since *PM*, pounding the 41-year-old Russell into the conviction, which overwhelmed him for several years, that he 'could not hope ever again to do fundamental work in philosophy'.

However, in the direct, near stenographic record of the 1939 lectures, Turing appears as an amiable, but persistent and vigorous foil for

Wittgenstein's explorations of mathematical proof, paradox and contradiction, calculation, and so on. Nearly every lecture contains one or two acute, staccato interchanges, often initiated by Wittgenstein, in which Turing on a few occasions scores quite sharply. Toward the end of lecture II, Wittgenstein abruptly calls on Turing to be a target for Wittgenstein's skepticism about referring to the infinity of natural numbers as a kind of number ('Well', responded Turing, wryly accepting his sacrificial role as conventional mathematician, 'if I were not *here*, I should say aleph-nought.')

From that point on, Wittgenstein increasingly wove the lectures into a response to Turing's claims and counterexamples, allowing, even provoking, Turing to respond, in some cases quite sharply and persistently, to all this remarkable behavior. Over half the lectures can only be described as an interlocked series of perhaps five intricate discussion/arguments between Wittgenstein and Turing, and while Wittgenstein does 85 per cent of the talking, Turing's lines are sharp, often decisive in that they, in a few sentences, set an agenda through a thesis that Turing often tellingly reiterates and strengthens as discussion proceeds, and while there is clear disagreement at points, there is much more a mutual sense that each calls forth the other's moves, and knows that his understanding is best tested and revealed against the other. Here is the harrower of the old paradigm setting tasks for the engineer of the new.

Both give the lectures; Turing's contributions are of a piece, in intricate use of concrete examples and in stark presentation of the argument, with Wittgenstein's more extended, more far-ranging, complementary analysis. Both are 'seeing one figure as the limiting case of the other'. Professors Norman Malcolm and John Wisdom, and others, enter the discussion peripherally on a few occasions. Characteristically, Turing wrote only, at the time, that Wittgenstein would cast him as spokesman for official mathematics and then, after suggesting what his views must be, proceeded to make gawky Turing feel a little foolish.

In any case, the captivating skeptical drive of Wittgenstein's later work was to *show up* the notions of meaning, thought, rule-following, language, and mind that lead to myth-ridden psychology, whether everyday, introspective, or behavioral, and that in fact led to his own mistaken views in *Tractatus*. These notions had also led to formalist projects like *PM*, whose elaborate theory of levels and types Wittgenstein had seen as a technical dodge, one doomed in principle long before Gödel's telling technical objection. Consequently, Gödel's demonstration simply added one more piece of counterevidence.

Since the roots of these notions are deeply embedded, and savingly veiled from view, in ordinary life, Wittgenstein is also doing a kind of phenomenological survey, or cognitive psychoanalysis, of the patchwork system of inhibitions that normally veils and smoothes us through the many incoherences and paradoxes that lurk near the surface of our everyday life. The introduction supposed that we 'quarantine' paradoxical sentences and, more generally, fail to see readings (or hearings and sayings) that lead us into nonsense. But *how* is this done? And how much are we repressing? Wittgenstein's answer to the second question is: a lot.

Since Wittgenstein saw his role as showing up myths and misunderstandings, he had no direct interest in helping construct what would follow this ground clearing.

Wittgenstein had, through much of his life, exhibited an interest in building things unusual for a philosopher or pure mathematician, though he hardly approached the prodigality of Turing. As a boy Wittgenstein most impressed his dazzlingly wealthy and talented Austrian family by making a sewing machine, so they sent him to a nonclassical secondary school and he then spent two years studying mechanical engineering. He came to Britain in 1908 when 19 to study aeronautical engineering for three years. After attending a term of Russell's lectures on mathematical logic, Wittgenstein asked Russell, 'Am I a complete idiot?', suggesting that were he not, he would become a philosopher. Russell, pleading a lack of data, proposed that Wittgenstein submit a philosophical paper to help Russell resolve the question and, after reading the first sentence, told Wittgenstein he must become a philosopher. Though Wittgenstein had suggested that aeronautical engineering would be the appropriate occupation had Russell judged him a complete idiot, idiocy would come to seem to him a state of purity one could attain after putting down the seductive, prideful illusions cast by mind and language.

That Wittgenstein designed and successfully built a house after writing the *Tractatus* illustrates the paradox with which he prefaced his book: that it essentially solved the problems of philosophy and in so doing made clear how little had been done in doing this. Wittgenstein saw constructive enterprises such as house-building as a rest from philosophy and logic. When he decided that the *Tractatus* was just one more illusion, Wittgenstein also came to see that there was no general, permanent solution to the illusions of mind and meaning conjured up by humans and their languages. The best he could do was collect loosely linked sketches in which he relentlessly used sharply etched descriptions of everyday human life, in which machine-like procedures are exemplified, to shine contrasting light on the legions of myths he wanted to dispel. If he inadvertently and brilliantly managed to show ways of delineating precisely the families of illusions, bred in everyday life, that threaten understanding of Turing's thesis and construction project, and cognitive science more generally – and indeed also brilliantly laid out the budget of problems that motivate the computational approach *and* pointed to a variety of paradoxical phenomena that have become crucial avenues for the cognitive sciences – all this would not have made him a happy recruit in the brave new constructive enterprise. For him engineering was consolation, not philosophy.

Philosophical Investigations begins with a quote from *Confessions* in which St Augustine relates how he learned language by correlating his elders' sounds and the objects they pointed to or grasped.[2] Wittgenstein later explained that his target was his own *Tractatus* view that *a word's meaning is the object it stands for* but he quoted St Augustine's similar, casual formulation to emphasize the depth and authority of the view. Wittgenstein responds to St Augustine

Now think of the following use of language: I send someone shopping. I give him a slip marked 'five red apples'. He takes the slip to the shopkeeper, who opens a drawer marked 'apples'; then he looks up the word 'red' in a table and finds a colour sample opposite it; then he says the series of cardinal numbers – I assume he knows them by heart – up to the word 'five' and for each number he takes an apple of the same colour as the sample out of the drawer. . . . It is in this and similar ways that one operates with words. . . . But how does he know where and how he is to look up the word 'red' and what he is to do with the word 'five'? . . . Well, I assume that he *acts* as I have described. Explanations come to an end somewhere. . . . But what is the meaning of the word 'five'? . . . No such thing was in question here, only how 'five' is used.

Here Wittgenstein hits us with some points that develop through the book. The most basic, of course, is that 'apples', 'red', and 'five' do *not* stand for objects (physical things, color quality, number): they initiate *formalized procedures*. 'Five red apples' initiates a chained sequence of these procedures. Words do not mean, they do.

Of course, Wittgenstein is finding the roots of formalized procedure in the everyday human world, so his rendition is informal. But note the relentless way in which every operational step gives us a *step by step physical procedure* and, where one might be tempted to insert 'he *recognized* that the word "apples" meant this *sort* of fruit', we get the marked drawer procedure; where we are tempted to suppose 'he knew "red" meant *red* and recognized those as instances', we get the word–color sample procedure; and where we might expect 'knowledge of number' we get a rote counting procedure. Finally, to the summary demand 'how does he know how to . . .', we have the reply, 'I assume he *acts* as I have described'. The machine (the entity that just does what it does, period) is unparadoxic bedrock. Or rather, since approximations of formalized procedures are the unit of understanding, this is the closest to bedrock we have got.

(There is no *absolute* claim, of course. He acts as I have described is just what we happen to have. There is surely an incredibly complex account to be given of the (unconscious) neurophysiological sensory and motor processing that compose the bedrock of 'opening the drawer marked "apples"' or 'looking up the word "red" in a table'. But this account is not available: there is no step by step procedure actually in view. It is just that in *this* situation, as in the ever so many others that Wittgenstein constructs, *what people do* is the explanatory floor. We are not *required* to suppose that ghostly inner 'meanings', 'recognitions', 'knowings that', have to be secretively paralleling 'justifying', 'explaining', and 'giving real meaning' to manifest public procedures. Further, if we accept these questions ('but how did he know, recognize it as, etc.?' and 'but what did he mean?, what rule was he *really* following?'), *then*, as the mathematical quest that Turing resolved showed at a very different level, it *will not* be bedrock and you *will* feel compelled to invent mental acts of meaning and following a rule that *cannot* be bedrock and will cause paradox and all sorts of misunderstandings eventually.)

Similarly, Wittgenstein makes us see, quite sharply, specific incoherences in Augustine's account. If a word simply names the thing it stands for, what about 'Water!', 'Away!', 'Ow!', 'Help!', 'Fine!', 'No!'? If Augustine learned word meanings by associating a sound sequence with the object his elder pointed at, you ought to be able to do this:

Point to a piece of paper. . . . And now point to its shape . . . now to its colour . . . now to its number (that sounds queer). . . . How did you do it? . . . You will say that you 'meant' a different thing each time you pointed. And if I ask how that is done, you will say you concentrated your attention on the colour, the shape, etc. But I ask again: how is that done?

Considering this, child Augustine certainly might have a hard time *learning* word meanings (that is, how is he to deduce, from the finger movement you make, that you mean the shape and not the color? – Or the roughness or smoothness, the level or tipped; that this points to one belonging to the tribe, one who is tall, one who is to be emulated, etc. Might, too, the word and pointing mean *take that one there, remember that . . . , beware of these Xs*, and so on).[3]

Further, Wittgenstein exhibits 'games' as a striking example that both further shows up Augustine, naturalizes formal procedure, and explains the episodic, winding, paradoxical, incompleteness of *Investigations*. From board games to football, ring-a-ring-a-rosie to playing catch,

solitaire to the Olympiad, it is hard to see more than a family of resemblances, a few typical examples and inbetweeners, but no clear essential
characteristic, rather a sprawl built up of various analogies and extensions
whose more disparate instances have nothing directly in common. So, as
happens with many other words, when I 'point' to some instances of
'game', there is no Xness whatsoever that belongs to all those I point to
and no others. (It illustrates my thesis that Wittgenstein's 'family of
resemblances' analysis has directly spawned a major body of fascinating
empirical studies that identify patterns of such concepts in everyday
cognition – 'furniture', for example, is like 'game'.)[4]

As his 'five red apples' beginning suggests, a formalized procedure is
Wittgenstein's archetypal unit of cognition (computation), and he, by a
wider and more various route, has reached Turing's conclusion that
language, truth, and meaning (or procedural rules, interpretation, and
interpreter) can always give rise to paradox. Wittgenstein also draws,
much more richly, Turing's conclusion that use (contribution as a cog
in any sort of computational procedure, whatever its supposed aim may
be) is fundamental, not 'meaning' or 'truth'. But instead of going on to
the 'general form of the machine and its computative varieties', Wittgenstein turns about to survey the variegated sprawl of actual and imagin-
able human procedures, showing how much is taken for granted, how
much can give rise to an endless variety of breakdown and nonsense,
among which the multistranded, grandiose illusion which culminated in
his *Tractatus* is just one more.

There is no general form of the paradox, the illusion or breakdown,
and for *us* no general or final solution. It is our natural history, a
characterizing feature of our way of life, which is one of formalizing, or
inevitably looking like we are formalizing, in an endless patchwork of
ways, including, at the peripheries of our life world, *PM* and other large-
scale contrivances. So Wittgenstein pulls and pushes us through and
against the rough sprawl, the half-hidden, illusion breeding, structures
and ruins of our human world. Since use and procedure are basic,
Wittgenstein's short paragraph groups *plus a human reader reading* are
uninterpreted machines that spin and bump one about, expelling the
reader, who is now more on guard against various seductive cognitive
illusions. This is why Wittgenstein's sentences are, so often, *orders* or
questions; why the paragraphs, mostly, do not end with any concluding
generalizations, nor did the book (explanations come to an end somewhere, there is a last house on the street), which simply stopped. This
is also why the literal sentences, the sharply etched, family portraits, *are*
the *Investigations*, and much more properly and legitimately so than

those of the *Tractatus*. Naturally, *Investigations* has uses – its paragraphs *do* things to people – but, shades of the *Tractatus*, it has no thesis.

Meaning as Computative Function: Procedures and Misprocedures

Wittgenstein's striking sketches, so often striking unmaskings of cognitive illusions, have led some to think he was a behaviorist, meaning by that someone who claims that *there are not any conscious, inner experiences at all*. But that is evidently false.

Indeed, Wittgenstein, again and again, parallels actual public procedures with actual introspectively pictured ones. When handed the slip marked *five red apples*, and after opening the apple drawer, a second shopkeeper *might*, rather than reaching for the physical chart, have brought to mind a chart with imagined letter sequences opposite imagined colors, which he scans until he finds the letters 'red' and scans across to the color sample, finally saying the cardinal number words from 'one' to 'five' and, at each word, picking apples of the same color as the sample in his imagination from the drawer. So, Wittgenstein shows us, of course we have an imagination and of course we can do some of the subprocedures (subcomputations) there as well: but note that doing it *there* does not make the procedure more reliable or more real, nor is the function of the procedure made clearer there.

What Wittgenstein finds illusory comes out if we imagine a third shopkeeper, one who proceeds just like the others except that, instead of using the public color sample chart like the first or working up one in his imagination like the second, this third shopkeeper, after opening the labeled drawer, simply picks red apples from the drawer, one for each cardinal he rotely pronounces. The illusion we must resist here is the supposition that the third shopkeeper must be doing the color sample routine only he does it so fast or so well that he hardly notices. No, says Wittgenstein, he just does what he does, there is just no checking procedure. For picking out red apples, this automaticity works out fine for human shopkeepers, though this might not be true if the slip calls for 'carmine yellow' or 'hunter green'.

Bedrock, as Wittgenstein forces us to see again and again, has to be what people actually *do*, with words, charts, pictures, etc., as all run smoothly in everyday life. Of course, there may indeed be some physiological causal mechanisms that we might someday uncover, but all that amounts to *here* is that he acts as he does. The subprocedures of the shopkeeper example are in clear view (to us and to the participants) and we can reflect on variations, increasing or lessening or changing the subprocedures.

In a richly varied, and deeper and more comprehensive way, Wittgenstein exposes and explodes what cognitive scientists today call 'the little man in the head' fallacy of 'theories' of, e.g., visual perception that, under scrutiny, amount to sketching how retinal stimuli could be projected on an inner screen – which just regresses the problem of how the whole human being *sees* back into the mythic little man (how do *his* eyes work?, surely he must have a third, still littler, man inside him?).

Of course, as Wittgenstein insists, this is not just a fallacy, nor just an artefact of psychological theorists. There is a little man in that we *do* on occasion make pictures and talk to ourselves in our imagination, and *this* inner theatre seems, at least in a grainy and undetailed fashion, to be structured and shaped somewhat as the public world we see with open eyes. Even the illusions of mind that have bedeviled philosophers and scientists are rooted in ordinary life (and so are a subject of cognitive science as well as a fallacy). And the first great discoveries about the mechanisms of the eye, which founded scientific physiology, at the same time proved a vital impulse for 'the little entire universe in the head' illusion with which 'the little man in the head' illusion is paradoxically intertwined in a creator/creation interdependency. Wittgenstein begins to roll back the paradox from the little man end.

Wittgenstein did as much as any man to exorcize the problematic that seemed unavoidable to the vast majority of philosophers and cognitive scientists even into the twentieth century, one powerfully begun in Descartes' skeptical argument that one could be certain one was thinking while still doubting one had a bodily existence. Descartes, as we have seen, concluded that mind and body were separate substances, instances of two realms, and he went on then seemingly to recover the physical world of objects and other minds he had temporarily doubted. Subsequent thinkers found he had not sufficiently appreciated the difficulty he uncovered. Henceforth, one *begins* with one's immediately experienced inner world and inquires how the outer world of objects and other minds might be inferred or constructed from one's experience.

It is a measure of change that one is *startled* by Russell's 1925 introduction to Lange's *History of Materialism*, in which La Mettrie is restored and given some priority, though only as a quirky contributor to a persistent curiosity of intellectual history, one vulgarly naive and long known to be clearly false. Russell, while not endorsing Lange's particular refutation, clearly takes it for granted that the skepticism Descartes powerfully introduced shows materialism to be hopelessly naive (since of course the 'physical world' is available to *us* only through our sensations, we know we have minds and can only suppose it is likely that there are bodies). Hence, Russell concludes that we can only 'regard materialism as a system of dogma set up to combat orthodox dogma', one that gives way to skepticism if one is rational, its only remaining proponents, 'certain American scientists [behaviorists] and certain politicians in Russia, because it is in those countries that traditional theology is still powerful.' Russell, through most of his life, expressed the common conviction of the philosophic and scientific mainstream in maintaining that there is no way to refute solipsism, no way to be sure any other thinkers exist, or indeed, anything else, like the hand that appears to be in front of my apparent face, or the rest of the physical world.

What was believed, and what rationalized the doubt, was that, presumably somewhat in the manner of *PM* respecting logic and mathematics, statements about the 'outer' physical world could be reduced to 'more basic' and much longer statements about 'inner' sensory experience (as statements about number were reduced to logical statements). A physical world statement is a helpful abbreviation of a massively long conjunction of sensory experience statements. A physical object statement which has no translation into sensory language is undecidable and hence meaningless (what is the auditory experience of a tree falling when there is no one to have the experience?).

So, in this view, our everyday languages appear 'superficially' to consist of physical object sentences but their 'underlying forms' would be massively long conjunctions of sensory experience sentences. And, so this compulsive view goes, these 'sensory experience sentences' must exist and be determinate *since the only possible meaning physical object statements could ever have would be in terms of our possible sensory experience*. The paradoxical rationalization comes out when we realize that this means, strictly, that the physical world only exists as a shadowy abbreviation of the real world of subjective experience, but that this *cannot be any sort of meaningful problem* for us since meaning only has to do with our inner sensory experience, the bedrock. (This 'bedrock' of course now includes everything I shall 'inwardly experience', everything I shall ever see or hear – as opposed to *infer from* or 'translate into from' my experience – so of course the vivid and magnificently detailed world I see with open eyes is just as thoroughly subjective and mental as the drab, vague, eyes-shut version.)

6

The Contradictions of the Public World

One can say 'I will, but my body does not obey me' – but not: 'My will does not obey me.' (Augustine)
But in the sense in which I cannot fail to will, I cannot try to will either.
And one might say: 'I can always will only inasmuch as I can never try to will.'
Doing itself seems not to have any volume of experience. It seems like an extensionless point, the point of a needle. This point seems to be the real agent. And the phenomenal happenings only to be a consequence of this acting . . .
I may recognize a genuine loving look, distinguish it from a pretended one. But I may be quite incapable of describing the difference. And this not because the languages I know have no words for it. For why not introduce new words? . . .
Our business is not to resolve a contradiction by means of a logico-mathematical discovery, but to get a clear view of the state of affairs before the contradiction is resolved. Entanglement in our own rules is what we want to understand . . . Philosophy does not tell us how language must be constructed to fulfil its purpose, in order to have such-and-such an effect on human beings. It only describes and in no way explains the use of signs.
Wittgenstein[1]

The Contradictions of the Private World

The logical positivist analysis can appear as a simple generalization of *PM* (and did so once to this sentence's writer). But there are major differences between the *PM* project and the fuller project of working out the rest of the 'logical construction of the world'.

The foundational logical language of *PM* has a small, explicit, largely familiar notation, which embodies clear cut notions, commonly employed by logicians and mathematicians, much of it going back to Aristotle. *PM* is an immense work of logical and mathematical formalization (though, naturally, portions of it look clumsy by twentieth-century standards).

Scores of its formalizations and claims continue to animate waxing and waning subliteratures in philosophy, logic, mathematics, and theoretical linguistics, and, more indirectly, in computer science and cognitive psychology. In the main, these subliteratures differ from the predominant primarily interpretive, philosophical, even cultural, subliteratures of the *Tractatus* in being concerned with the truth, effectiveness, and application of various claims, demonstrations, distinctions, and formalizations that happen, coincidentally one might say, to turn up in *PM*. Appropriately, '*PM*' can refer to either the first printed edition or the second, which significantly simplified the solution to set theory paradox but at major theoretical cost, *or* to the two much longer, unwritten, but precisely specified texts that *PM*'s printed version greatly abbreviates.

Gödel pays tribute to *PM*'s inescapable centrality by making it the subject, via his system of unique numbering, of his enormously more prolix, literally unwritten but mathematically, precisely described, paradox-inducing construction. But, significantly, it is a fifth, unwritten *PM* (unabbreviated *and* missing the English text) that Gödel shows he can number. And then it is not so much that text alone, but the formal system specified by the text's formation and proof rules *and* the natural number system that formal system purports precisely to 'unabbreviate'. All of this becomes the *means* for Gödel's incompleteness constructions *and* the exemplary *subject* of his surprising results. What makes this demonstration possible is that, given a vast, systematic body of formally expressed mathematical knowledge, and clear standards for formalization and proof procedure, *PM* gives a precise description of a formal logical language, and a precise construction and demonstration, from this logical foundation, of the formulae of this body of mathematical knowledge.

On the other hand, the atomic proposition language that physical object statements supposedly abbreviate has never received anything remotely like this explication, even in the most loose and sketchy sense. Even 'physical object statement' is a deeply disputed and contentious notion, and there has hardly been even a sketch of a formalization of this notion clear enough to provide a meaningful target for 'logical constructions out of sense data'. 'Red patch here now' (understood solely as a claim about what was directly present in one's own inner experience) was offered with minor variations in countless papers and books as a lone rough suggestive example of what was intended.

The seemingly inescapable argument could be put in one sentence. Since we obviously have sensory experiences of this sort and since, indeed, such experiences are all we know directly and the only possible sources of meaning and evidence for what we find ourselves saying about the external physical world, *there must be such experiences and somehow*

*a language that describes them, for otherwise our physical object state-
ments would have no way to be true or false, no way to mean anything
to us.*

Wittgenstein's own *Tractatus* did not even give anything like an example of either an atomic proposition or the atomic fact it pictured. Quite
properly, a defender would say, because logic can only specify what these
must achieve, which is that they give determinate sense to the complex
'abbreviated' statements of empirical science. How, we might ask, do we
know that the literal statements of actual science actually have such a
reduction? The then compelling but empty answer is that only those
parts of them that have such a reduction are meaningful, are science.

When, indeed, Russell came, in the late 1940s and 1950s, to find
philosophical significance in recent discoveries of experimental science,
he himself, curiously, experienced the charge of materialist naiveté.

Lexicographer Samuel Johnson's naive way to refute skepticism about
the physical world was to kick a stone, saying 'I refute you thus.' An
Alan Turing might similarly propose to show a machine can think by
building one that actually does the trick (of passing as an intelligent
human through endless conversational interchanges). But this is the
outside way, one that will not work with someone deeply caught up in a
line of thought that makes Johnson appear naive and Turing and his ilk,
the would-be 'Promethean' toymakers La Mettrie predicted, equally so,
if more time consuming and elaborate in their supposed counterexample.
Frivolous, counters the sophisticated divine, where the forthright St Paul
would have said, or done, something far more forceful. 'Simply refuses
to see the problem', said several generations of philosophers and scientists,
secure in their Cartesian doubt, and in the two certainties that at least
followed, namely, that *there are thoughts, mental experience exists* and
materialism is self-evidently false.

(Language use *is* getting weirdly inverted here, for someone who says,
unlike Johnson's oafish gaffe, 'Is this a dagger I see before my face?'
should be asking a self-evidently most rational of questions, just the sort
of thing with which all serious rational inquiry begins, along with 'Is the
world around me just a dream?', 'Is a malignant demon distorting my
mental experience?', 'Is it not quite possible that I am the only consciousness in the universe?', and so on, just the sort of stuff Shakespearian
characters, worked to a pitch of tragic frenzy by the machinery of the
first four acts, start speculating about in the tortured epiphanies of the
fifth.)

For someone deeply caught, as Wittgenstein himself had been caught,
in an illusion, you have, somehow, to work your refutation from within.
Wittgenstein was surely the right man for the job.

When, at the beginning his *Meditations*, Descartes claims, with a pro forma expression of great bewilderment, that he has realized he has no absolutely conclusive evidence that he is awake and that his sensory experiences are not completely illusory, we have some sense of rehearsal. This impression is reinforced by the comment with which Descartes prefaces his studied surprise, namely, that the doubt, which he is amazed to see is so unavoidable, is one whose statement might heretofore only be expected from a madman in an asylum. Freud's claim that paranoia is a caricature of philosophy, just as compulsion neurosis is *also* a caricature of religion, can be read in various ways, one at least of which is that Cartesian doubt is a professional tool or ritual, an impersonal, collective maneuver that purports to achieve a certain collective result through sophisticated verbal maneuver in which ordinary words like 'doubt' drift far from usual employment ('the language is idling' Wittgenstein would say – *Investigations* scourges such idleness as cognitive sin, the entry for evil beasts both impish and all-encompassing).

Russell tells us that he sat baffled by set theoretic paradoxes for two years, feeling quite sure that there must be some simple technical solution, feeling that the problem was trivial and he, slightly ridiculous. Though *PM* and related works forged the structures, techniques, and rationalizations for seeing language and formalization as bedrock, Russell never fully joined the revolutionary shift in which the *Tractatus* triumphs. Through his long life, Russell never quite accepted Wittgenstein's claim, reiterated by decades of logicians and physical and cognitive scientists (particularly, the logical positivists of the Vienna Circle), that logical and mathematical 'truths' are the analytic upshot of linguistic conventions (most narrowly, the 'tautologies' of the *Tractatus*' mechanically decidable propositional logic).[2]

Yet Russell also saw, and did more than anyone to make others see, that the phenomenal empiricism of Locke and Hume required that there be, somehow, a sense data language, one that is private and logically uncommitted about the physical world or other minds. It was this that motivated the *Tractatus* and, in making phenomenal empiricism explicit, led it into the demolition, the naturalizing of formalization, of *Investigations*.

But perhaps since Russell could not fully commit himself to the 'linguistic turn', he could not take the internal collapse of the *Tractatus* project, and related projects, sufficiently seriously to see the point of the passionate scourging of Wittgenstein's later work. And, more importantly, there is the difference in intensity and Paulinity. Recalling Descartes' pretense of startlement that he cannot tell whether he is, or is not, dreaming (a claim, he had just written, that we think a clear sign of a madness), perhaps one needs someone who embraces a sophistic, illusion-breeding argument so deadly seriously that he exposes its pathological roots. Russell, in introducing his *Lectures on Logical Atomism* (1919), which express views like those of *Tractatus*, remarks, with disarming breeziness, that a philosopher begins with premises that appear so obviously true that no one could doubt them, and then proceeds to deduce from them, step by undeniable step, conclusions so wildly absurd as to be totally unbelievable. Wittgenstein viscerally revolted at such talk.

Russell gives a witty account which illustrates the contrast. He tells us that Wittgenstein, who arrived at Russell's Trinity College rooms at midnight, announced that he was going to commit suicide immediately after leaving. As Russell puts it, he did not want, given Wittgenstein's announcement, to hurry him off. Wittgenstein sat in silence until three o'clock, when Russell asked, 'Wittgenstein, are you brooding about philosophy or your sins?' to which Wittgenstein replied, 'Both.' (In his *Autobiography*, and in letters, Russell says, without embellishment, that Wittgenstein came to brood suicidally most nights for several weeks; two of Wittgenstein's talented brothers had committed suicide in his adolescence, and more were to do so.)

Earnest, determined incredulity, a refusal to accept the disarming reassurances of Descartes and Russell, is not enough to put down a well-established multilayered illusion, for one can now easily dispense with its cruder beginnings (like a ladder you climb up, and commanding a clear view, now see was not really there). G. E. Moore, whose relationship with Russell began when both were Cambridge University undergraduates, earnestly and intensely questioned the consistency and authenticity of thinkers who would 'doubt', or even 'deny', the most basic facts of the ordinary world, while at the same time comfortably function and speak within that world. 'Does *that mean* you *did not* have your breakfast *before* your lunch???' Moore earnestly and intimidatingly demanded of professors who 'proved' time was unreal and the like.

In a lecture, Moore also gave a Johnsonian proof that physical objects existed by raising his hand before his face, claiming that this was certainly a hand, and then repeating the sequence with his other hand, finally concluding that physical objects existed (he added the gloss that he was *so sure* of his two premises that he felt no counterargument could have premises quite as sure). Moore's argument, though clearly in the Johnsonian vein, makes a crucial concession that Samuel Johnson did not. Moore suggests not that he just plainly *sees* his hand (thus *knowing* it, even without the Johnsonian flourish of *kicking* it), but more that he is *so driven* to the view that he has hands it is wholly unlikely that a counterargument's premises could ever overpower this compulsion. Note that he still leaves open the possibility, for him unbelievable and doubtless very improbable, that he does not have hands.

Moore, in fact, when asked what he meant by physical objects was inclined to say that, while he was quite sure that they existed, the problem of the *analysis* of what 'physical object', or 'is a hand', *meant* was complicated and somewhat unclear. Indeed, Moore made it evident, in other writings, that he held much the same view as the tradition: that we *see* sense data, not physical objects, and that 'physical objects' are something like 'logical constructions out of sense data' (i.e., that is what 'physical objects' *really* are, and so, indeed, not something we see, or even infer, but something we have *constructed* as an abbreviation of hordes of interacting sensory experiences). Skepticism about the existence of physical objects is smoothly absorbed into a project of analysis; like the Cheshire Cat's smile, or a *Tractatus* ladder, what the doubt conjured up still remains in Moore's conviction that talk of physical objects ultimately has meaning only through its analysis into propositions about sense data.

Another confirmation of the illusion's grip is that Russell, perhaps the most acute, prodigal, and relentlessly logical thinker of the twentieth century, should have presented, as decisive, the following argument: *'naive realism' gives rise to science, science refutes naive realism (by, e.g., Descartes' investigation of the mechanical properties of the eye), therefore naive realism must be false.* But the argument could just as well go: if science is true, then I do not ever *see* my hand in front of my face, etc., etc.; but I *do* see my hand in front of my face, etc., etc., and therefore, science is false. Or rather, the interpretation made by Russell of what Descartes uncovered when he dissected eyes is false.

Descartes showed us a mechanical eye, an information processing device, and Descartes sketched the extension of this account to other senses and to motor responses, indeed to all that we shared with the most advanced beast machines, thus for Russell seemingly pushing *consciousness and its contents* ever inward. But an even more plausible reaction, one whose fearful logic, even without La Mettrie's eventual explicit statement, called forth the passionate attack on the beast machine hypothesis, might be to understand it all – sense organs, nerves, and brain – as a physical, information processing system.

Wittgenstein will have none of it. Wittgenstein's internal refutation of solipsist mentalism begins neutrally, so to speak: we shall examine the experienced texture of everyday life, including you and me, without deciding such sweeping questions. We first notice that, typically, the meaning of a word or sentence (idea or thought) is use, its step by step contribution to a procedure. The various incoherences found when we try to take up the view that words *mean* in simply naming things (and sentences mean in picturing facts), also confirm this procedural approach. Procedures can often be undertaken in both the public and private world (surely often less reliably in the private world, as suggested by the second shopkeeper's replacement of the first's color sample board by an imagined one).

Wittgenstein provides prodigal, multifaceted examples of how clearly our cognitive life largely operates in the public world (cognition belongs to the 'public big man', on occasions supplemented by imagination or talk within). When we write, or draw, we think with our pen, just as we measure with rulers, reckon with sextants, calculate with pencil on paper, compose with our typewriter, understand the words we speak in hearing them, paint the subject's likeness with a brush, count with our fingers, move wooden chess pieces by the rules (thus threatening, tricking, attempting), and so on. The everyday world is full of mind and meaning. It *is* the use, the cog-like place within the workable procedure, that invests words, tools, actions, sentences, and subprocedures of all sorts, with meaning, and their users with intention. Though many particular procedures have to be public, it does not matter in general whether they go on publicly or privately; to the degree they are anywhere, meaning and conscious cognition are in the public world.

However, in order to keep in mind this obvious feature of everyday life, one has to guard against the many temptations to supplement, fill out, justify, and ground the 'rough, slapdash, *ad hoc*, gappy, everyday procedures' not by actually adding or doing anything specific, but by acceding piecemeal, and eventually whole hog, to the illusion that all this is somehow *already* taken care of *inside* our mind. So Wittgenstein shows us, again and again, variations of the shopkeeper lesson that of course you *can* add particular procedural checks, or switch from public to private or (usually a practical improvement) the reverse. *But* checks like explanations trail off somewhere and there is not the slightest hope of procedures to safeguard against all possible complaints. *And* we must resist the illusion of procedureless procedures.

Illusory procedures are ones that are *not* open to view, ones without visible steps, ones that suggest no way of telling whether they have actually been followed step by step or whether some steps have been skipped, procedures of unknown numbers of steps that (for all we shall ever know) occur instantaneously and unwilled, and surely might not be there at all. 'The language is idling', he would say, and he harrowed such useless dirt, such cognitive sin, with relentless fervor. Naturally, 'know' was particularly scourged.

In our everyday world, we often say or think 'that looks (sounds, feels, appears to be, etc.) like a such-and-so' and we may then undertake some specific checking procedure we have developed, and so come to say appropriately, 'now I know, it is one for sure, I checked, etc.'. No idling here. But the third shopkeeper, the one who simply picks out the *red* apples, one for each rote count cardinal up to 'five', does not he have to 'know' or 'recognize' them to be *red*? Wittgenstein's answer is 'he just picks the *red* ones out' and that is all there is to the procedure. If you want to recommend adding something to the procedure (like the color chart, for example), fine, and *then* we shall be able to say whether he has checked or not. (Of course, there may well be neurophysiological mechanisms, but those are not step by step procedures, open to view and checkable: in *our* terms, he simply acts as he does.)

Similarly, if you, to check on legibility, show me a letter you have just written, and ask 'Could you, please, read this?' and I reply 'Yes, I shall try' and then proceed to read aloud the entire letter (you hear my words and they match what you wrote), and I now comment 'Yes, there, I read it'; if you now ask, 'But do you *know* you read it?', I might reply 'I *did* read it' or 'What do you have in mind?' If I do say, nettled, 'Of course I know I read it', I am really just rejecting the suggestion that some further, more thorough, check is needed or available; I am not, certainly, asserting that, in fact, I went through some additional practical check, which I did not reveal to you (as a powerful, spottily dyslexic billionaire, I have my room bugged and hidden zoom-lensed video cameras patrol it to keep a precise record of what I do, say or write; through a microphone in my ear, my henchlings told me my reading performance checked out perfectly).

Wittgenstein strongly objected to frivolously and uselessly adding 'I know' to simple straightforward statements about what a person sees, thinks, or feels when no further check is sensible or even possible, e.g. 'I have a toothache', 'Here is a large oak', 'I am thinking of a cardinal number from one to five'. Feeling pain lancing through your jaw, can you then, through some special act of concentration 'know' that this *really* is pain? One particularly scourges here because if you allow the idle 'know' in here (and in all similar cases), then the way is open, down the road, for someone, noting that there, indeed, *is not* any further checking procedure that justifies the addition of 'know', which given its idleness there will not be, to point out sternly that we really do not know, or that 'we only know in our own case', or finally, 'only I know!' And so: on to solipsism!

Wittgenstein's attack on solipsism is thought by some to climax in the so-called 'private language argument'.[3]

I want to keep a diary about the recurrence of a certain sensation. To this end I associate it with the sign 'S' and write this sign in a calendar for every day on which I have the sensation. [A definition of 'S' 'cannot be formulated': it is just *this* sort: but what sort is that?, is there an ostensive definition?] Can I point to the sensation? Not in the ordinary sense. But I speak, or write the sign down, and at the same time I concentrate my attention on the sensation – and so point to it inwardly. . . . I have no criterion . . . whatever is going to seem right to me is right . . . 'Well, I *believe* that this is the sensation S again.' – Perhaps you *believe* that you believe it!

Wittgenstein goes on for some paragraphs about the idleness here. What is the *use* supposed to be? What is identifying it again *for*? Those who co-opt Wittgenstein as defender of the life world against scientific pretensions, claim that here Wittgenstein makes most sharply the central thrust of *Investigations* in a conclusive argument against a 'private language'. So they would say something like,

In forming a language for describing his private experience, he would never be able to tell whether S had occurred again, so he could not describe his private world to himself; so, there can only be *public* language, language that presupposes other persons and everyday experience, and solipsism is impossible. The public life world of everyday experience is finally made secure, restored from the sophistical arguments begun in Cartesian skepticism. The world of ordinary life is restored. That is Wittgenstein's fundamental achievement.

That *Investigations* begins with a quote from St Augustine (AD 354–430) and that most of it shows incoherence and paradox in *public* procedures might make one skeptical of this assessment. That Wittgenstein intended *Investigations* to make anyone, least of all humanist professors, comfortable and secure is absurd.

Wittgenstein's actual lines clearly tell against solipsism, of the specific *Tractatus* form, in which the 'object world language' intertranslates with the 'sense data language'. This is of course the logically secure and comfortable solipsism of Russell and Moore, the one that says in effect 'since object world descriptions are just abbreviations of more foundational, but much much longer, sense data descriptions, our worry about whether there are, or are not, objects there when we do not look (or mysterious I-know-not-whats) is meaningless'. We have got our cake and we have eaten it too – solipsism without pain.

But this line means that I have to have a sense data language out of which to 'logically construct' the object world. Wittgenstein's argument is that this demands that it be possible to characterize this sense data world completely independently of the object world (which, supposedly, *is just* an abbreviation). And this characterization procedure is bogus: indeed the very idea of *checking* procedures is missing here.

Yes, Wittgenstein makes clear, you effectively *can* use an imagined copy of a public world color sample and color-word table. But that would not work, Wittgenstein points out, with an imagined table that had no relationship to our public world. And why not? Because there is no procedure. I can check my imagined color table against the public one from time to time, but no checking procedure is available for the *purely* private table.

The Contradictions of the Public World

Well, we can surely say, that certainly does put us into the public world to some degree and by default. But remember that much of what Wittgenstein says, in casting doubt on this private language, he has already said about the public one and he goes on with this critique, in multifaceted ways, through the rest of *Investigations*.

In the logically-private language case, Wittgenstein says he could 'point to it inwardly' but that this does him no good because he does not have any criterion for *what* (what identification procedure or use) he is pointing to (he can 'concentrate his attention on' some indescribable feature as opposed to others, but that does not do anything). But *that is child Augustine's problem to a tee*: what is the elder pointing at (and what is the point of this whole pointing procedure anyhow?). And the reason that this situation *is public* is simply that how the elder acts, the physical things about me, and how I act *are* at this public world level *independent*. Child Augustine will learn from (be surprised by) the wallops of his elders and the world. How much, however, will he learn and how much will work because his human internal mechanisms make him generalize in the same ways as his elders? How much will this public world have a determinate, coherent, sense (without assumptions about unknown mechanisms)? How much will *it* be foundational bedrock, self-contained on its own terms? Is it perhaps more a case of you can be in my dream if I can be in yours, of necessarily mutual illusion? Recent linguistics begins in the demonstration that children cannot acquire their first language unless they are programmed biologically to do this, to make the right generalizations without any evidence or teaching.

Recent commentators have centered their attention on the bleak skepticism revealed in Wittgenstein's much longer public world discussion of public world cases of 'following a rule' (i.e., domestic variations of formalized procedures). In his *Investigations* and his lectures on mathematics and elsewhere, Wittgenstein goes on and on about numerical versions of this. Suppose I start X counting 2, 4, 6, 8, etc., and he says 'I know how to go on' or 'I know the rule'. When he reaches 100, however, he continues 104, 108, 112, and I complain that he is mistaken, he is not following the rule. Startled, he says that *of course* he is following the rule. This might even be the case if I say, or write, some formulation in English or algebraic notation: 'But that is what the formula *says*,' he remarks in frustration at our complaint. Wittgenstein also tries to show us that I may in such cases have to realize that I had not in some absolute sense specified the interpretation when I produced the formula.

He gives a variety of related cases: difficulties in saying when and which games people are playing (since *anything* they do could be described in a complex of rules); in pointing out how many incompatible ways are inevitably available in trying to describe what game(s) people are playing in throwing a ball about and running here and there; trying to figure out the language of a newly discovered tribe, a task, Wittgenstein tellingly remarks, for which we properly make 'the common behavior of mankind' a 'system of reference', and yet still may fail; and, finally, difficulties in the experience of rule-following (the match between some symbols on the blackboard of my imagination and what I do is one target for destruction), where the paradoxes of interpretation multiply cancerously, and where it becomes patent that talk of the rule *causing* my behavior does *me* no good. This final subjective discussion terminates with something very much like Turing's resolution of Gödel's paradox:

When I obey a rule, I do not choose. I obey the rule *blindly*. [I do it *automatically*. And the symbolic expression which I did not *choose* to follow:] My symbolic expression was really a mythological description of the use of a rule.

We most want to say we are following a rule when we move from input to output *automatically* (blindly) like someone counting or adding numbers on a bill. But there is absolutely no actual experience of 'consulting' or 'looking at' *the rule*, rather my finger moves and my mouth calls out 'one', 'two', . . . , or I see numbers appear on the page at the end of my pencil. *Why should it go this way?* . . . Wittgenstein, we remember, is describing the everyday world from within. Only procedures that are open to clear view count. No hokey occult unobserved procedures allowed. But what becomes clearer and clearer is how starkly inadequate 'what is open to view' is to give anything like the assurance, the determination, that we habitually assume our everyday procedures to have: and paradoxes and visual illusions erode our sense that 'what is open to view' is so, is bedrock. What seems clear is that the experienced world, in being so, requires a physical and computational world beyond.

It is, of course, no part of Wittgenstein's project to say anything about this, for the point of his enterprise, as he announces at many points, is not even to suggest explanatory theories but simply to take a close look at what is open to view in our ordinary world, to expose features that are inexplicable or paradoxical at that level:

If the information of concepts can be explained by facts of nature, should we not be interested, not in grammar, but rather in that nature which is the basis of grammar? . . .But our interest does not fall back upon these possible causes of the formation of concepts; we are not doing natural science.

Precomputational behaviorists like Russell's 'American scientists', experimental behaviorists in the James Watson–B. F. Skinner tradition, linguists like Leonard Bloomfield and Benjamin Lee Whorf, and similarly inclined anthropologists and philosophers, might say this:

What Wittgenstein overdramatizes is that the physical object world makes linguistic meaning more *indeterminate* than expected by those swaddled in a prescientific language and culture. That is why Professor Quine uses the 'radical translation' frame, in which the anthropologist only observes regularities in the native's physical activity within the physical neighborhood. All that is determinate about meaning are correlations between a rabbit bustling through in clear view, the anthropologist's reproduction of a previously observed native sound sequence *gavagai*, and the native's assent [*gavagai* = 'rabbit', 'it rabbits' etc.]. Quine's point is that you do not get any *more* than this; you do not get the native's metaphysical theories about rabbits, which are not genuinely meaningful. In his remarks about pointing, knowing which aspect, and 'following a rule', Wittgenstein finds *only* that physical correlations make meaning far more indeterminate than what mentalist illusions require. And that is all to the good!

When we reach the chapter on the new computational linguistics, we shall let a Wittgensteinian-spirited linguist further skeptically harrow Quine's little story. Recent linguistic science begins with the task unmasked by Wittgenstein's critique of Augustine's commonsense view of language learning: what are the unconscious, innate mechanisms that shape language acquisition?

Yes, the meaning part of Wittgenstein's 'life world' *is* fragmentary and much more indeterminate in meaning than we think until we take a careful look (and we do not get comforting vagueness but paradoxically conflicting demands). But, Wittgenstein shows in a series of remarkable examples, that our world also suffers from paradoxical overdetermination. Wittgenstein points out that we *see* quite differently: a written word upside down, the same graphic configuration right side up, and both instances in a wholly unfamiliar notation. When someone speaks *our* language, we hardly hear the actual sounds, while, reading print, we seem still to see the marks. What we might like to think of as observed sounds or marks are shaped through *computational reflexes* so to speak – mostly we just cannot see or hear them in a way that brushes away the effects of our individual linguistic experience (surely the innocuous so-called 'physical sound sequence', *gavagai*, might not only sound *very different* to anthropologist and native, but the native might distinguish *gavagai* so very differently from adjoining phonological structures that the anthropologist would simply misidentify much of the time, so the correlation procedure cannot get started).

Apart from the paradoxical way that language-bearing sounds and marks look, if we human beings are *really seeing* the *real* world then it has got to look *the same* or anyhow *roughly the same* when we look at the same thing. If it is not, the everyday cognitive world is not cognitive bedrock, for its supposedly physical objects are woven, and woven differently, by the different perceptual/acquisitional development of its inhabitants. Wittgenstein makes similar points about faces and facial expressions.

Concluding sections of the *Investigations* discuss visual paradoxes. For example, the duck-rabbit figure, which humans can see as one or the other without in either case the suspicion that any alternative exists, can also shuttle from one to the other (like many visual paradoxes), *but* one must see it as a duck *or* as a rabbit, and never as a neutral, inbetween smear. We seem forced *by cognitive reflex* to see it as a duck, now as a rabbit, but never anything in between. But it is not the marks, impossibly physically changing, nor some lens distortion, but *my* cognitive activity must be doing this *and that is not part of my life world*.

I am shewn a picture-rabbit and asked what it is; I say 'It is a rabbit'. Not 'Now it is a rabbit'. I am reporting my perception: just as if I had said 'I see a red circle over there.' . . . But how is it possible to *see* an object according to an *interpretation*?

When it looks as if there is no room for such a form between other ones you have to look for it in another dimension. If there is no room here, there *is* room in another dimension.

This is like the third shopkeeper, who just picks out the red apples, only there is no option of replacing automaticity (he acts as he does) with a step by step, openly followed procedure. We *have to* see it as a rabbit or a duck, and there is no way for us to be conscious of the visual computations that give rise to this dichotomy. If what we see was really just a flat pattern on our eyes *or* the actual physical world, this *could not* happen: there is not a duck-rabbit in the world but only in the paradoxes of our visual perception of it.

Wittgenstein's 'private language' argument against solipsism is simply that unless I admit the existence of *independent procedural checks*, I cannot in some literal sense check my own meaning. To have to admit more: one, another, on up to the everyday world of things and people, might be called a 'relative inconsistency proof' in that the solipsist world does not work on its own, relative to the everyday world, so the everyday world is relative to *that* more consistent. Does this everyday life world make coherent sense on its own terms, so *meaning* and *thought*, and *language* and *picturing*, as well as robust folk psychological procedures and families of expressions, make a coherent, stably determinate, self-contained world *here* and *now*, one that stands in no need of further cognitive explanation? – No.

With inhibition-unmasking sketches, Wittgenstein pictures our fragmented, parochial, cacophonous, paradoxical, multipurposed, illusion-breeding, yet unavoidable, zoo of procedures, partial procedures, illusory procedures, paradoxical procedures, and their parts, such as words, sentences, figures, pictures, facial expressions, rule-following, seeing-an-aspect, intending, feeling, counting, eyeing, hearing – parts paradoxically inadequate to their role, parts that work (courtesy mother cognitive nature) much better than they ought, parts whose shape we see plainly as Dr Johnson's stone, yet know may at the next moment flip into another configuration, parts that almost seem designed as illusory ladders for the prideful I.

A behaviorist like Quine programmatically *takes it for granted* that his spare, physical object world is unproblematic, and then uses indeterminancy to clean off mentalist cobwebs and rationalist dogma, thus triumphantly leaving and validating, as intended, the extentional meaning and logic, and empiricist psychology, that the behaviorist wants.

The 'ordinary language' philosopher or psychologist wants a rather richer, more literary, more idiosyncratic, life world. Yet he, like the behaviorist, *also* wants the same assumption of self-contained meaning, so as to make the life world a self-sufficient foundation from which to disparage various targets – cognitive psychology, materialism, rationalism, mentalism, even science in general – in so far as they purport to have anything to say about *meaning*, *thought*, *consciousness*, seeing, hearing, intending, and indeed, the whole variegated vocabulary and correlative textures of the life world. One imagines them saying, 'Having closed the illusory gap opened by Descartes' anatomy lesson and *Meditations*, we are safe, and intend to continue to be safe, from the mumbojumbo claims of would-be "cognitive scientists", safe in our comfortable, self-contained, age-old, life world.'

Only Wittgenstein wants *to look and see*, *to examine close up*, to start with the phenomena of everyday procedures (or comparative inventions) and let this examination naturally reveal the degree of coherence, self-containedness, and meaning-sufficiency in the life world. In this sense, Wittgenstein alone takes the actual phenomena of language (in the larger context of procedure) as his foundational starting point. And Wittgenstein finds that the life world lacks coherence and self-sufficiency, though, of course, it is, and is in part *by that paradoxical token*, the only world the I has got: You and I have to believe in it. By virtue of its resembling-families sprawl of partially formalized, often paradoxical and illusion-breeding, procedures and procedural parts (signs, symbols, samples, experiences and perceptions of such and others, etc.), *the life world is intranslatable and hence irreducible*.

PM and arithmetic, Turing Machines and computability, as well as substantial portions of empirical science, and languages with spoken and written versions, codes and ciphers, do intertranslate and, in some cases, we find some explanatory reduction. The respects in which, as Wittgenstein often says, a Martian could not understand us, nor we a lion, are those for sure that ensure that the works (programs) of Plato and Paul, of Shakespeare, Dickens, and Kafka – and *Investigations* – will still retain the *use* they have had for us, whatever we may discover, as cognitive scientists, about the mechanisms of the computational mind.

But cannot I imagine that the people around me are mere automata, lack consciousness, even though they behave in the same way as usual? – If I imagine it now – alone in my room – I see people with fixed looks (as in a trance) going about their business – the idea is perhaps a little uncanny. But just try to keep hold of this idea in the midst of your ordinary intercourse with others, in the street, say! Say to yourself, for example: 'The children over there are mere automata; all their liveliness is mere automatism.' And you will either find these words becoming quite meaningless; or you will produce in yourself some kind of uncanny feeling or something of the sort.

Seeing a living human being as an automaton is analogous to seeing one figure as a limiting case of another; the cross-pieces of a window as a swastika, for example.

Juxtapose this concluding remark from *Investigations I* with the 'five-red-apple' procedural examplar that opens the book.

We have our firmest sense of bedrock, of language *use*, when it is all running on automatic ('the machine is running smoothly'), when questions of meaning, of rule-following, of who is doing what and how *do not arise*. This is also so at the level of individual experience: when I run on automatic, *others* (and I, unless I think about it) find my cognitive performance unquestionable. – And even more so for the ever unconscious part of the operation, the part that computes, from retinal irradiations, three-dimensional representations within us, and from sonic vibrations, words and sentence structures, that not only seems to function rapidly and automatically, with quite daunting computational tasks, but whose computational output both draws the familiar, experienced bedrock of the life world and, interlocking with other computational modules, determines state changes, stores data, and initiates, and coordinates, motor activity. Behold the computational mind.

It is the conscious mind and the 'public big man' that we shall see as the limiting case of the causal/foundational computational mind and the large-brained social mammal, the creator side of the creator/creation paradox.

Yet, as Wittgenstein's 'but just try to keep hold of this idea in the midst of your ordinary intercourse with others' reminds us, a kind of cognitive reflex makes it quite impossible for us to hold on to this view in everyday life, respecting others or oneself (even, for the most part, respecting mammals or computers). Indeed, 'I see people with fixed looks (as in a trance) going about their business' juxtaposed with 'the children's liveliness is mere automatism' quite rightly suggest that, within the everyday world, where everything is open to view, we have not *any* general procedure (indeed, not even a hint of one) for deciding, of any particular human, whether he/she *is* or *is not* a computational machine (any biophysiologist will tell you that liveliness is a much easier muscular orchestration problem than fixity mixed with slow, uniformly timed movements). So the *cognitive reflex* expresses a normally unavoidable attitude, something analogous to a duck-rabbit figure which we almost always see as, say, a duck and only when we meditate strangely does it flip momentarily into the rabbit.

In the next chapter we shall meditate strangely with Alan Turing on what might convince us, what test procedure might establish, that an electronic digital computer *thinks*. While Wittgenstein will make a final appearance, in dialogue with Turing, at that chapter's beginning, Wittgenstein, as I have remarked, had no interest in becoming an empirical scientist, nor in describing or building computational machines.[4]

A proud young man, aflame with the grand illusions of Pure Intellect, wrote (*Tractatus* 5.62), 'That the world is *my* world, shows itself in the fact that the limits of the language (*the* language which I understand) mean the limits of *my* world' and, labeling himself 'the philosophical I, the metaphysical subject', authored the ultimate, complete account of all the stark minimal truths to be had about Language, Mind, Meaning, Logic, Thought, Object, I, and Fact.

The man who said the world was his comes now to say 'five red apples', as one, abjuring illusionary pride and the vaunting ambition of the metaphysical I, who shall talk plain, who shall confess, who shall abase proud reason, ridicule and harrow it to its roots in everyday humble nonsense, routing reason's pretentions even in the grocery store.

The criteria for the truth of the *confession* that I thought such-and-such are not the criteria for a true *description* of a process. And the importance of the true confession does not reside in its being a correct and certain report of a process. It resides rather in the special consequences which can be drawn from a confession whose truth is guaranteed by the special criteria of *truthfulness*.

Wittgenstein takes *truthfulness* as an important narrative virtue, requisite to confessions. *Investigations* is the metaphysical subject's confession. Wittgenstein's apology had already been written by W. B. Yeats:

> Those masterful images because complete
> Grew in pure mind, but out of what began?
> A mound of refuse or the sweepings of a street,
> Old kettles, old bottles, and a broken can,
> Old iron, old bones, old rags, that raving slut
> Who keeps the till. Now that my ladder's gone,
> I must lie down where all the ladders start,
> In the foul rag-and-bone shop of the heart.[5]

7
Turing and Wittgenstein

T: *You have a system of calculations, which you use to build bridges. You give this system to your clerks and they build a bridge with it and the bridge falls down. You then find a contradiction in the system. – Or suppose two systems, one of which has been used satisfactorily for building bridges. The other system is used and the bridge falls down.*

W: *I am a general. I tell Rhees to be at Trumpington at 3:00 and at Grantchester at 3:30, and I tell Turing to join Rhees at Grantchester at 3:00. They compare orders and find 'That is quite impossible.' Given a certain training, if I give you a contradiction, you do not know what to do. In logic, we think of the calculations and ways of thinking we do in fact have – the technique of language we all know. Here contradictions do not normally occur – or at least occur in such restricted fields (e.g., the Liar) that we may say: If that is logic, it does not contain any contradictions worth talking about.*

T: *If one takes Frege's symbolism and gives someone the technique of multiplying in it, then by using a Russell paradox he could get a wrong multiplication.*

W: *Suppose I convince Rhees of the paradox of the Liar, and he says, 'I lie, therefore I do not lie, therefore I lie and I do not lie, therefore we have a contradiction, therefore $2 \times 2 = 369$.' Well, we should not call this 'multiplication'; that is all.*

T: *If there are contradictions, it will go wrong.*

W: *Nothing has gone wrong that way yet. Why not?*

Though the interchange, in the 1939 lectures, goes on to a discussion of intuitionism and mathematical truth, this leg of the discussion compactly pictures the dialectic of a contrasting yet complementary Wittgenstein and Turing. Wittgenstein relentlessly insists that we force ourselves to see how indeterminate, patchy, and contradictory our actual symbol use is, how much its successful operation must depend on hidden peculiarities of human cognition nature.

As ever, Wittgenstein sees himself as simply describing our actual cognitive/linguistic practices. As in other natural languages, English grammar seems to allow us to construct Liar paradox sentences; and Frege, for example, put together a formal system in which Russell constructed a paradox. Neither practice seems to have got us into all that much trouble: there are far worse incoherences in our everyday lives, ones which are actually productive of major illusion and misunderstanding. If we come upon a Liar paradox, we often simply ignore it, rule it out of our calculations, but what we *do not* do is go on in the way jokingly imagined for Rhees. Similarly, when Rhees and Turing discover that the general's orders are contradictory, they will not rush about every which way generating general anarchy – no, they will simply ask for further orders. We must see, Wittgenstein stresses, that 'a contradiction is not a germ which shows general illness'. *We* are innoculated somehow in a way that computers sometimes are not.

Descriptively, Wittgenstein is right about all of this, though the *New York Times* recently reported 30,000 new viral infections in our computer offspring. Frege surely felt discomforted when he read Russell's postcard but he did not wander about compulsively spouting paradoxical deductions like Rhees – Frege, the interpreter, thus somehow caught in cogs of his formal invention.

Turing's position is that he does not want the bridge to fall down. In application, a contradictory calculus might produce a mistaken calculation and this in turn disaster. Further, it may produce this result without the contradiction having to appear in the glaring manner that Wittgenstein imagines.

Wittgenstein points out that all sorts of calculating slips cause trouble. Why, then, so much fear of a mere possibility that *might* arise from a contradiction, indeed a possibility perhaps never in fact realized, which, since meaning is *actual use*, may suggest that contradiction is not quite what we take it to be. Wittgenstein, the anthropologist of sign use behavior, suggests that 'what they do when they get to the contradictions will depend on what reasons they had for holding to the formula – on how much the formula *means* to them'. Yet Wittgenstein also, eventually, concedes that we, as a *practical* matter, avoid contradictions, but this is more practice and human nature than pure logic and mathematics (similarly, the X who continues the sequence 2, 4, 6 . . . 98, 100, with 104, 108, etc., has an unusually odd and inconvenient nature, and it is fortunate that this is extraordinarily rare, perhaps even bioneurologically impossible, among us humans). While, similarly, the normal child Augustine *naturally* took up his normal elders' patterns of generalization, though neither he, nor his elders, could have said *how* this was done, *how* humans unconsciously sort speech sounds, gestures, objects, and aspects so that language acquisition is possible.

W: . . . If the rules for obeying these orders – for logical product and negation – are laid down, then '*p* and not *p*' cannot make sense and we cannot obey it.

T: I should say that if one teaches people to carry out orders of the form '*p* and not *q*' then the most natural thing to do when ordered '*p* and not *p*' is to be dissatisfied with anything which is done.

W: I entirely agree. But . . . does 'natural' mean 'mathematically natural'?

T: No.

W: Exactly. It is not mathematically determined what is the natural thing to do . . . An explanation of *why* a contradiction does not work is always just another way of saying that we do not want it to work.

In this final, at least seemingly sophistical, comment, Wittgenstein appears to play again his point that the 'private language' speaker, confronted with what seems to be *S*, cannot distinguish between recognizing this to be *S* and choosing to extend '*S*' to label this sensation, too. In Wittgenstein's life world logic, no new step in symbolic manipulation is automatic, every step is a choice. So if Turing's clerks build a faulty bridge, this must be a 'mistaken application'.

So far, perhaps, so even. Turing wants to build reliable bridges. Wittgenstein wants to describe our actual (conscious, intentional) symbolic practices.

If, however, one wants to build (or describe/design computationally), not a bridge, but a computing machine, if one wants, as Turing several times announced in the early 1940s, 'to build a brain', *then* matters are not so balanced, *then*, as our experience with computer hardware and software has so amply and revealingly shown, a contradiction may not only reveal general illness but may be the germ whose cancerous multiplication causes the general illness. If, in the spirit of Turing's 1937 paper, we build the interpreter and interpretation *into* the formal system, then it *will* crash.

And – turning back on ourselves – this practical computational travail reinforces Wittgenstein's taunting question about human cognition: 'Nothing has gone wrong that way yet. Why not?' Wittgenstein, given his descriptive methodology, actually and revealingly, had no way of explaining this obvious feature of our cognitive experience. The problem was one he uncovered, along with so many others, in tracing the peculiarities and paradoxes of our conscious cognitive existence. Relentlessly, he insists that there is nothing 'in clear view', in learned, consciously available, human *conventions* that guards us against disaster; Wittgenstein entirely agrees that human nature somehow saves us but reiterates that this is not our formal *conventions*. I suggested, in my introduction, that the Liar paradox once briefly made me queasy. Why did I not, like a computer, *crash*? Why do humans, in general, not *crash* when they run into such sentences? How are we protected?

MIT's Marvin Minsky, whom the *New York Times* called 'the father of Artificial Intelligence', found Sigmund Freud's works in some boxes one of his graduate students had abandoned. This accident led Minsky to generalize Freud's dirty joke theory.

Freud maintained that we unconsciously inhibit our recognition of sexual or excremental 'dirty' words, or interpretations of words that make them so: the punch line produces the tension-releasing laugh by tapping the emotional force of the inhibition, by 'rubbing our noses in it' through an unexpected turn of phrase ('evil beasts, swollen bellies').

Several experimental studies confirm that perceptual inhibition is quite real. In one, subjects tried to recognize words flashed on a screen for various brief intervals. 'Dirty' words took substantially longer for recognition and misguesses about such words (but not about 'clean' words) often changed *every* letter, suggesting not partial recognition but recognition and suppression (e.g., 'totals' guessed for *breast*, rather than 'brother' for *bother*). Abrupt changes in galvanic skin response, blood pressure, and heart beat suggest that the 'dirty' word is unconsciously identified in less than half the time required for conscious recognition of 'clean' words – but *then* some inhibitive reaction misleads and delays conscious recognition.

Professor Minsky, round whom grew up the first generations of programmers and 'hackers', knew intimately two salient features of computer programming: 1) Computer languages strive to avoid all forms of ambiguity; 2) If there is ambiguity in language or program, if there is some bizarre, unintended interpretation, the computer surely will produce it, for serendipitous weal on occasion but nearly always for woe – and if there is a paradoxical order, the dutiful machine may loop endlessly or spume up contradictory nonsense and crash. 'How come we are so different?' so one imagines him adding to Wittgenstein's 'Nothing has gone wrong that way yet. And why not?'

Unlike Minsky and Turing, who would want to understand how human cognition does, and machine intelligence can, avoid falling down, Wittgenstein does not wait for an answer, for there is no answer at the level of everyday lived experience. Epimenides and the righteous Paul so long ago noted the evil beasts. Logician Russell, mathematician Tarski, and countless formalists, insisted that a proper language must grammatically exclude paradoxical sentences and its semantics must not produce contradictions, for otherwise disaster must occur. Wittgenstein relentlessly insists that natural human languages apparently allow paradoxical constructions and semantic contradictions, and yet the 'inevitable disaster' does not occur (it did not even occur with, e.g., Frege's unrestricted formal language).

Imagine the following story. Suppose Frege starts actually thinking and speaking in his own logical language (before Russell's telegram arrives). Well, then, would he not go strange-loop crazy, babbling in inferential circles? Surely, not! But *how* is this avoided? Deduction is the only clear cut formal operation that looks like thinking: you have a body of data (postulates) and inference rules, and you work out what follows from the data; there are sorting and halting problems of various kinds, of course.

Since anything true of a formal language ought to be explicable in natural language, one seems stuck in a variation of the circular paradox of common sense and 'science' that Russell claimed Descartes' anatomy lessons exposed. In formal language we model basic logical features of natural language, but we find that when we translate Liar sentences into our formal language, we then, formally speaking, can prove any absurdity whatsoever, and hence the Liar sentences have to be ruled out in a corrected grammar of intended-to-be-consistent logical language: *So*, natural language must be inconsistent. On the other hand: in some formalizations of the logical and grammatical features of natural language, natural language is inconsistent, *But* natural language *is* consistent (if anything can be so for our cognition), *so* those formalisms are themselves mistaken.

Thus, destructively and mistakenly, one might argue *à la* Russell that natural language leads to formal language and formal language shows that natural language is self-contradictory and false, therefore, natural language is self-contradictory and false. Equally, since natural language seems to work all right and since formal language suggests it should not, there must be something wrong with our formalizations, something left out or over-simplified. Positively, while builder Turing must consciously employ a consistent and simple design calculus, the student of human language and cognition must recognize that since natural language, at the descriptive level available in ordinary conscious experience, contains contradictions and hence should crash but does not, *there must be levels of description and processing, not available in ordinary experience, that explain how we are protected.* Levels and processing that will, more generally, explain why, in normal conscious speech processing, we some-how unconsciously filter out the many mishearings and misreadings that ambiguity allows.

Here we see the strange harmony between Wittgenstein and Turing. Wittgenstein relentlessly impunes the comfortable assumptions of Russell,

Tarski, *et al.*, who see paradoxical natural language sentence paradoxes as destructive viruses, as formal structures that a viable language must rule out, but who fail to see that since the destructive epidemic has not in fact occurred, *the picture they give of the natural language Liar sentences, OR of natural language sentence processing, is misleading.* So, positively speaking, Turing sees two, intertwining, projects of cognitive science: 1) to build and cognitively employ electronic brains that have adequate formal power, speed, and reliability to embody efficiently and viably the symbol-manipulating capacity of Universal Turing Machines; 2) to structure and program such machines to simulate, approximate, embody, and so explain human cognitive and symbol-manipulating activity, assuming, e.g., that natural language is not in practice paradoxically self-destructive but rather, in syntax, semantics, or application, finesses or blinkers paradoxical collapse. And the Wittgenstein/Turing moral is as always that we ultimately can only assign form and meaning to sentences within a full characterization of its users' natural cognitive system in its neurologically, developmentally, and socially mature form.[1]

(When I first ran into an electronic calculator, a large and impressive 1950s business machine, I immediately keyed it to divide two by zero and happily sat back to watch it dutifully struggle along at an endless task, likening my 13-year-old self to Bulfinch's Greek gods, who sentenced Sisyphus to unending rock rolling. After 20 minutes and some increasingly worried button pressing, I felt more like the sorcerer's apprentice who is learning that a little knowledge of conjuring can be prodigally dangerous. Hearing the approach of authority, I finally pulled the electrical plug from the outlet and as rapidly replaced it. I later learned the machine had a red button on its back, whose sole purpose was to stop such a process. Having recovered my spunk, I was disappointed that the next electronic calculator I ran into, cunningly, lacked the relevant zero button. In more recent decades programming and programs have evolved ever more sophisticated protections against crashes of quite various sorts, with virus protection now almost an industry and, not unnaturally, also instructively opening a way of modeling and understanding the DNA viruses that have evolved biologically. While nature may not positively plug pull, the possum's protective nervous collapse, 'playing dead' in face of immediate, solutionless danger is a red-button style solution. Our everyday response to paradoxical sentences suggests higher level programming protection and this seems unavoidable in explaining how we avoid the host of ambiguity problems that Marvin Minsky points out. Critics warn us that nature, Richard Dawkins' *Blind Watchmaker*, continuously spawns new solutions out of old materials and that we should be silly to think that its solutions will have the elegance and simplicity of our own, from scratch designs. But if one looks to actual evolution of machine intelligence in the last several decades of the twentieth century, one finds all the make-do, accident, and adaptation and readaption of older materials and systems to newer environments that nature practices.)

When we are building, we look for possible design language paradoxes, to root them out so the thinker will not misthink, the mechanism, misbehave. When we are looking at human language and visual perception, we look for possible paradoxes as a way of finding out peculiarities of natural design, clues to underlying structures and processes, questions not defects.

Aristotle might classify the first undertaking as practical engineering, the second as theoretical cognitive psychology and brain science. Yet, in fact, the first, artificial enterprise has, over the past 50 years, made major contributions to the second, theoretical investigation. Indeed the two undertakings are inextricably intertwined. Cognitive science necessarily embraces, and must flower if it can, in this uncertain interplay. May Oscar Wilde's benign paradox carry the day.

8

Information Storms

The fact Babbage's Analytic Engine was to be entirely mechanical will help us to rid ourselves of a superstition. Importance is often attached to the fact that modern digital computers are electrical, and that the nervous system also is electrical. Since Babbage's machine was not electrical, and since in a sense all digital computers are in a sense equivalent, we see that this use of electricity cannot be of theoretical importance. Of course electricity usually comes in where fast signalling is concerned, so that it is not surprising that we find it in both these connections. In the nervous system chemical phenomena are at least as important as electrical. In certain computers the storage system is mainly acoustic. The feature of using electricity is thus seen to be only a very superficial similarity. If we wish to find such similarities we should look rather for mathematical analogies of function.

Turing[1]

The excellence of reason does not depend on a big word devoid of meaning ('immateriality'), but on the force, extent, and perspicuity of reason itself. Thus a 'soul of clay' which should discover, at one glance, as it were, the relations and consequences of an infinite number of ideas hard to understand, would evidently be preferable to a foolish and stupid soul, though it were composed of the most precious elements. A man is not a philosopher if, like Pliny, he blushes over the wretchedness of our origin. What seems vile is here the most precious of things, and seems to be the object of nature's highest art and most elaborate care. . . . But since all the faculties of the soul depend to such a degree on the proper organization of the brain and of the whole body, that apparently they are but the organization itself, the soul is clearly an enlightened machine.

Julien de La Mettrie

Wittgenstein could not leave his private language diarist, or his private language, alone. His conclusion of the example epitomizes the turn to cognitive science. He imagines he discovers one day that when he has a sensation that he guesses to be, or calls, 'S', 'a manometer shews that my blood pressure rises'. Wittgenstein finds this a 'useful result', for he now can tell when his blood pressure rises without any apparatus. If he now suspects that one or two of the Ss did not have quite the right inner feel, it will not matter, assuming he continues to recognize when his blood pressure rises. Objects pull. An 'S' recorded in his diary will mean that the blood pressure in his brain has increased; his experience has become a perception of this rise.

Wittgenstein showed us how much the everyday world we experience wears our cognition on its sleeve, but further and more specifically how in linguistic and visual forms, we see paradoxical impossibilities in the world in front of us that can only be the upshot of clashing information-processing procedures within our brains. The dim, wavering world within has sense only as it echoes the procedures and processes of the public one without, which has its own incompleteness and paradox, suggestive of nature and neurology beyond clear view. Subjective inner states are inevitably connected with public criteria, though these in turn often rest on neurological and physical factors beyond or below view. In his discussion of pain, as a uniquely experienced subjective state, Wittgenstein insisted that a sufficiency of public behavior meets the criterion for saying someone really is in pain, though we of course cannot feel his pain. Blood pressure, however, is a physical condition inside the brain. While the manometer technology provides a reliable public way of determining this inner physical condition, the diarist might now be said to have learned an equally reliable private way to perceive these changes inside his brain. The example has more general application: robust oddities in linguistic and visual perceptual experience may converge with sophisticated technological measures in determining higher level brain function.

The pull of everyday objects and actions, which works cognition and subjective experience into clear view in the everyday world in which we experience our lives, and so allows a sharper, more objective view, can also come to be exerted by physiological states and processes in the brain, even those dedicated to cognition. After a time, the diarist may simply say, 'Oh, oh, there it goes again, my blood pressure is up' and perhaps have no S experience at all any longer, just the vague impression that his brain is overly pressed with blood. Wittgenstein, indeed, anticipated the use of feedback to help patients to sense and often exercise some rough conscious control over heart beat, blood pressure, etc. A lengthy effort is required, driven by the health use, and the patient has only the vaguest sense, if any, of what the physiological condition is like or what controlling it is like (control is hard to achieve for blood pressure as opposed to heart beat and body temperature). Technology, manometer or other feedback, provides an informative link to help forge these humble twinges of self-knowledge.

Deaf people can acquire a much more solid and definite sense of, e.g., the wall looming up in front of them, a flatness ahead, not leftward or rightward or, still less, behind or above. Background sounds, reflected informatively from the wall, are the actual source from which the perceived flatness is constructed. However, the deaf person denies *hearing* anything (hearing consciously), and he denies rightly and usefully, for it is the spatial information, projected through a most native and powerful visual/spatial faculty, that is meaningful and insistent. You cannot dodge around, or bang your forehead into, *sounds*. Interestingly, technology can give a better picture by providing an input that is literally pictorial. Bearing a forward-focused video camera, whose input image is transmitted electrically into a similar, larger pattern of stimulation delivered to the nerves distributed over the skin of a blind person's back, initially provides a pattern of back itches, which come by fits and starts to be interpreted as clues to a world in front, until, after several days of use, the back itches may be felt no more, disappearing into a patchy and somewhat dim, three-dimensional perception of the world ahead.

These possibilities of direct self-knowledge, or better, object-driven self-transformation are quite restricted. Over a vast range of native spatial and linguistic perception, no sort of scientific knowledge or demonstration *will change what we see or hear*. The native cognitive contribution is robust, reliably replicable, like the reflex your physician engages by tapping your patella – knowledge of kneecap reflexing has no practical influence, the leg robustly bounds up at the tap whether the patient understands or no, and the physician expects the same were she to tap her own patella. Indeed, perhaps the most substantial, systematic and detailed advances in cognitive science of the last few decades have been made in linguistics and visual perception, where natively acquired cognitive systems perform robustly and stably whether the subject's *conscious, theoretical knowledge* of them is copious, minuscule, or nonexistent, accurate, mistaken, or rapidly changing. You might complain that *know thyself* ought to be a more practical imperative; self-knowledge surely should work more change and be more personal, even perhaps more profitable.

Plato insisted that Socrates' imperative leads, above all, to knowledge of what is universal and unchanging, indeed to what is seeded within already, programmed to sprout. In modern linguistics the point may be made nonmetaphorically and with scientific precision. The normal human naturally acquires powerful and detailed tacit knowledge of her or his native language. Here is Mettrie's most 'enlightened machine', indeed the 'object of nature's highest art and most elaborate care', exhibiting in action 'the force, extent, and perspicuity of reason' speaking and hearing and understanding hundreds of new sentences daily from an infinite stock, operating with a speed and fluency grounded in a cognitive organization that constitutes the species, honed to full competency by decades of continual employment. Linguistics aims at theoretical knowledge, description and explanation of this marvellous competence of this 'clay' machine, whose functional organization is the essense of human cognition, the high road of the computational mind.

Turing's vaunting, long-term ambition for cognitive science was to construct, given the seemingly incredible speed and massive memory of the electronic computer, some approximation of this competency. We next turn to this ambition and how, in the information hurricane's eye of the Second World War, he came to it. His ambition may only be fulfilled given decades more of massive effort, research, and ingenuity, and it may well prove to require processing speed beyond the capacity of known computer architecture, if we leave aside our own neurological wetware (this is now the actual case for an activity thought ideal for computer success, chess playing: Hi-Tech, a leading mainframe chess program, *is not fast enough* to beat the top human players under the tournament requirement of 30 moves in two hours; if it were eight hours, Hi-Tech might well be unbeatable).

To illustrate simple, preliminary points about programming possibilities, Turing often described and worked with 'paper machines' in which Turing himself would be the computer, following out the very restricted rules of a simple programming language through the very many steps required for quite simple calculations. In fact, he made many mistakes and usually did little more than a few runs, just to see whether the technique looked promising enough to merit further examination on the much much faster and more reliable electronic computer. In effect, he was sketching a way of transcribing what a human seems to achieve in a small number of complexly coordinated steps into a very very long number of much simpler steps, the simpler steps, indeed, that were available in his spartan programming language. Translations of this sort *inevitably make the machine input hard for humans to use and impossibly long and tedious in computational requirements*: compared to our cognitive equipment, Turing always emphasized, *speed* is the only advantage the practical digital electronic computer has to overcome its many deficiencies. So whatever wisdom is gained in computer simulation and theoretical knowledge, it may not so much *improve* our native cognitive skills as *replace or enhance* them in the way that alphabet technology did, for example, by providing a common, reliable, artificial memory, etc.

Indeed, given all Turing's ways of converting complex input into much longer, simpler input, and the same with output and internal processing, we know *in principle* that there has to be a Turing Machine that represents the computational specs for my mind/brain (my memories, my talents, my brain's way of regulating heart and lung activity, everything) – or yours, of course. But there is no *literal* real time existence for that Machine, which will, e.g., translate into several million operations your vision of the ball rocketing along, off the foot of one of the divergently moving players, all built up and transformed via unconscious computations, into your automatic foot placements and kick. In the *in principle sense, you are that Turing Machine, but you do it all through a marvel of cunningly coordinated parallel processes, shortcuts piled within short-cuts, so densely structured that it befuddles the human observer.* You are, as Turing put it, a practical computing machine, one that meets the theoretical spec.s in a timely and reliable way.

The question Turing soon confronted was whether, and how, one could give the structure of a human mind *somewhere well short of* just plain describing its entire structure down to the electrochemical potentials of each nerve cell and the totality of their interconnections. Obviously, an in principle solution to the chess-playing problem would be to give a complete neurological simulation of world champion Gary Kasparov's brain; if you run the simulation program you will get his level of play and thus, at least, stand even with the best human player. The point is that while duplicating Kasparov's brain is surely guaranteed to do the job, the simulation will of course duplicate vast hordes of Kasparov's other or accidental capacities, features, and memories, and it will inevitably use exponentially much more neurological structure than some minimum needed for chess playing. Since neurological tissue provides the most compact information storage and processing medium we know of, any attempt to simulate fully any human brain at the neurological, or bottom, level will carry memory and processing requirements far beyond anything we can envision for future computing machines. The lesson of near 50 years of work on chess-playing machines is that just the job of functionally simulating Kasparov's chess capacity requires stupendous quantities of memory, heuristics, and processing speed.

Turing's Race

Syntax is the study of the principles and processes by which sentences are constructed in particular languages. Syntactical investigation of a given language has as its goal the construction of a grammar that can be viewed as a device of some sort for producing the sentences of the language under analysis.

Noam Chomsky[2]

In 1937, Alan Turing's boyhood interest in codes and code-breaking flowered electromagnetic-mechanically at Princeton University, where a fellowship had brought him to talk about undecidability with Alonzo Church and John von Neumann. Electromagnetic relay switches, then long in use in telegraphy, react to a pulse of electrical current by clicking *shut* if open or *open* if shut. Turing recognized that here were the flipflops which could play the role of the Turing Machine tape's 1 or 0, and its other minimal operations, provided they were wired together in branching logical trees. A physics graduate student helped Turing exemplify the discipline confounding character of cognitive science

Turing actually designed an electric multiplier and built the first three or four stages to see if it could be made to work. He needed relay-operated switches which, not being commercially available at the time, he built himself. My small contribution was to lend Turing my key to the Physics Department's workshop, which was probably against all regulations . . . He machined and wound the relays; and to our surprise and delight the calculator worked.[3]

Turing recognized that *speed* was the essential cipher consideration. A really good cipher would be one that would 'take 100 Germans working eight hours a day on desk calculators 100 years to decipher by routine search'. Also opting for speed electricomechanically, German military transmissions were automatically enciphered through an enhanced version of the commercially available Enigma Machine, the radio transmission then to be deciphered by an identical machine.

On 4 September 1939, three days after Hitler invaded Poland, Turing reported to Bletchley Park, where he was soon to build and rebuild networks of relay switches (whose portentous clicking and enigmatic exteriors led large units to be called, with a party spirit, not *bombs* but 'bombes'). Turing also crafted the heuristic software that eventually enabled the *bombes* to provide the plain texts of German naval communications within hours for much of the war, with ups and downs as the Germans made mistakes or changed procedures. In 1944, further processing speed was gained by replacing the clicking relay flipflops with the purely electrical vacuum tubes of *Colossus*. Turing held that code-breaking, along with chess playing and theorem-proving, would provide a good field for the beginnings of machine intelligence work. Let us see how the problem set Turing by Enigma led him to this view.

Speaking strictly, a *cipher* is a *set of rules* that transforms the plain text letters of a message into others, which the intended recipient usually makes sense of by reversing the rules. For millennia we have used fixed-number, addition ciphers: e.g. A (plus 5) is F, B is G, C is H . . . [looping around] W is B, X is C . . . ; and so, PROCEED HOME is

UWTHJJI MTRJ (the reverse, minus 5, deciphers). [Again speaking strictly and by contrast, a *code* works with larger plain text units such as *words*, rather than letters.]

The would-be decipherer of UWTHJJI MTRJ has many clues, which will grow if the message is longer. The repeated plain text letter that ciphers into JJ cannot be W, X, Y, Z, Q, or H because these letters do not appear doubled (in conventional English). It is unlikely to be A or U. *Conditionally*, if it is B, D, F, G, L, P, R, S, or T, *then* the next plain text letter, as a matter of English phonology, is going to be a vowel and if not Y, then I, beginning the sequence ING. Since the plain text doubled letter must alphabetically directly precede or follow one such vowel, we exclude all but B, D, F, or P, and, since such consonants, when doubled in plain text English, also have to be preceded by a vowel but there is no such vowel alphabetically available given our assumptions, we are stuck with the vowels O and E as the final candidates for the JJ double. Since the rule that converts JJ into OO makes garbage of the rest of the text, we are left with EE and the cipher rule must be ADD FIVE – and we have the message.

Note what makes cipher-breaking so easy in this simple fixed-number addition case. Obviously, English phonological rules play a major role. In longer texts, syntactical rules would also play such a role (some regularities would belong to all natural human language but it is obviously an enormous help to know what the language of the plain text happens to be). That favorite of literary cipher-breakers like Edgar Allan Poe and Doyle's Sherlock Holmes, statistical linguistic features such as letter frequency, can also play a part. E has greatest frequency, followed by T, etc., so one simple cipher-breaking technique is to assign E to the most common enciphered message letter and T to the next, and then see if the result is consistent with English and, if so, check out further guesses, and if not so, reverse the E and T assignment and check it out, and so on and so on. Guessing and understanding might also depend on semantic meaning and pragmatic context. PROCEED is a stereotypical British naval command and so might be something to try out for one of the first words of a message. And we can only locate the referent of HOME if we know what boat is addressed and its home port.

What leaps out at once is that the grammatical rules of English, like any natural language, *powerfully constrain* what letters (phonemes) or words can appear at any point in a sequence. In letter sequences, there is very often only one possibility, so there is no choice and the letter chosen provides no (new) *information*: it is *redundant* in the quantitative information theory that Claude Shannon shared when Turing briefly did liaison work and engineering at Bell Lab in 1944. The interwoven redundancy of natural language is quite helpful in that we can initially miss or mishear many speech sounds in a sentence and yet grasp the sentence perfectly.

Moreover, the structure of the redundancy is a product of – and hence for cognitive science a key to – underlying linguistic and cognitive structure and processing. This we shall see prospectively exemplified in toy-like miniature in Enigma Machine deciphering, which suggestively came to involve a functional simulation, and modular differentiation of, the settings of the Machine's internal rotors, rings, and plugboard. Turing's effort would depend on both the natural, largely unavoidable redundancies (i.e. highly constrained structure) of the German language *and* redundancies produced by the arbitrary, abruptly changing peculiarities of German software and hardware settings and additions, plus the basic physical structure of the Machine.

From the encipherer's view, natural language redundancy *does* ward off miscommunication. If one enciphers and transmits lists of numbers that are for the receiver completely random (and so, with *no* redundancy), the receiver can have no clue that a mistake has occurred. Similarly, the most powerful cipher-breaking has *no* chance of deciphering such transmissions *even if* the cipher rule is our risible, fixed-number, ADD-FIVE rule.

Nonredundancy in the enciphering rule can have equally strong effects. What is called the 'one-time pad' – pages of random numbers shared by sender and recipient which sender adds, looping through the alphabet, to encipher the plain text letters one by one, using each add-number only once – provides cipher that is unbreakable for a third party (because it is much too cumbersome, the one-time pad technique is employed only for short, super secret diplomatic communication).

But though the encipherer can see safeguards as well as risks in natural language redundancy, and such natural redundancy is practically unavoidable in any case, *much can be done to increase exponentially the difficulty in finding the redundancies in the enciphering rule*. The Enigma Machine provided a relatively simple electromechanical way to produce rapid and reliable enciphering and deciphering for a vast horde of daily message traffic while seemingly rendering it practically unbreakable.

The basic machine had a 26 letter alphabetic keyboard electrically connected with three (later five) rotors, each with 26 positions, arranged like those in a speedometer so that the first could rotate through 26 positions before moving the second rotor one position, and that second rotor in turn would eventually move the third, so that there were 17,576 possible rotor states.

Thus, having set the rotors to one of these states, the encipherer would type the first plain text letter and the consequent electrical impulse, transformed in letter with each rotor contact, would terminate in lighting a particular letter. The machine state changed automatically, the encipherer wrote down the lighted letter, and then repeated the process, thus enciphering the second message letter via a new machine state (or new cipher rule) and so on. Note that the double letter of our single, fixed-number addition cipher would *not* appear as a double letter in the enciphered radio broadcast because the machine state, or rule, would change after the first letter of the double was enciphered.

How would the intended recipient decipher the radio message? The receiver's Enigma Machine would start deciphering with exactly the same rotor settings as the sender, so that E would uncipher back to A, the receiver machine moving automatically into the same second state as the sender machine, ready to uncipher the next message letter back to the plain text second letter, and so and so on, until the full plain text is retrieved (the receiver Enigma ending in the same state as the sender at message's end).

Obviously it is extremely convenient to have one rotor setting (and one standard machine) for both enciphering and deciphering. Indeed, linguistics shows us nature exploiting the same *inversive* efficiency: the same rule system comes into play in both *hearing* and in *speaking*, an inversive relationship which cognitive scientists often label 'encoding' and 'decoding' respectively. Thus, when we successfully *con*verse, we at some level of abstraction share functionally equivalent 'Enigma Machines' with the same settings. Our broadcast, the physical sound sequence of an English sentence, is a thought encoded by speaker and decoded by hearer.

But the efficiency the Germans gained created a *machine* redundancy, a structural fingerprint, along with the whole rotor system, that Turing exploited. For example, no letter would ever be enciphered into itself and, more generally, logical patterns were created which often allowed the cipher-breaker to exclude large numbers of possible setting solutions. To the commercially available Enigma, the German military attached *rings* to each rotor that added an additional alphabetic fixed-rule letter resetting, and a plugboard which arbitrarily exchanged, by rewiring, the letters going in and coming out of the rotor system. *These* additions, while reassuring the military mind and exponentially increasing the number of machine states, did not eliminate the rotor 'fingerprints'. Turing and his colleagues were able to distinguish the characteristic redundancies produced by these *modules* from each other and from those of the rotors.

(From the 1950s on, cognitive psychologists have labeled their work as 'black box' investigations. Since the neurological 'wiring' is much too complicated and miniaturized to make sense of directly, the psychologist

seeks to determine what minimal functional innards produce the characteristic, discriminating *outputs* of the human mind to experimentally
varied *inputs*. Some cognitive psychologists have taken to saying that
they investigate the software of the hardware brain. Still more recently
cognitive scientists have profitably added *modularity*: the search for the
characteristic contribution of separable 'black boxes' or mental faculties.
Again, artificial intelligence aims to simulate human cognition on an
electronic digital computer, ultimately on chains of flipflops, while at the
level of a programming language you get something on your monitor that
looks like letters, words, and sentences, something that looks a lot like a
human language because these light patterns arranged in these sequences
on the monitor screen are what our highly particular human species can
most easily understand and make use of. But the simulation is not just
a convenient way, so to speak, to state clearly and compactly what we
have observed and recorded about human behavior. The supposition is
that this properly programmed computer will now embody something
like our functional innards: the programmed computer will not simply
describe thinking, it will *do* it.)

Had large numbers of transmissions been enciphered on the same rotor
and plugboard wirings, and the same rotor and ring settings, cipher-
breaking would not have been enormously difficult. In fact, rotor wirings
and positions, and plugboard wirings, were occasionally changed, ring
positions changed daily, and each enciphering clerk would select a new
rotor setting for each message. This individual message setting, however,
would then have to be transmitted at the beginning of the message by
means of a three-letter *indicator* (rather like the operation instruction,
followed by input data, of a Turing Machine). In 1935, German naval
cipher clerks would put the machine on the day's *ground setting* and then
each would transmit the triple letter of their individual message choice
twice; finally, each would set the rotors to his individual choice and
transmit his message.

The Poles would concentrate on the first six letters of every transmission, seeking what common *ground setting* would cipher *all three-
letter sequences, repeated twice* into the vast variety of six-letter sequences
they had collected. With a catalogue of 106,446 relevant rotor positions
they were able to determine the day's ground setting within 20 minutes
and so decipher the individual message settings and messages at will
(recall that since the Machine would change internal state after each
letter-enciphering, the plain text repeated triple would *never* reappear,
after enciphering, as three letters repeated twice). In 1938, the cipher
clerk changed to giving a further *individual* ground setting in the message's first three letters, and *then* used that to cipher the repeated triple

which set the cipher for the rest of the message. This meant that the correspondences in the collected repeated triples would now have to be filtered through the 17,576 variations given initially in order, finally, to determine the *common* initial ground setting. When to this the Germans added a change from three to four rotors and various other changes, the Polish effort collapsed.

Aside from a massive increase in automated electromagnetic relay boards, which in effect *simulated* the functional innards of the Enigma Machines, Turing added a systematic use of conditional implications and consistencies as if a portion of *PM* were flexing its muscles. Turing extended this technique beyond the search for indicators to provisional word guesses and other searches for regularities throughout the message. The system he and his colleagues put together depended less and less on precise weaknesses of a particular indicator system, which could change, producing blackout, at any time. More and more, the total pattern of communication and events, through varieties of machines for different services, and map coordinate systems and known movements of ships and supplies, all fed the Bletchley effort, so that it came to provide a continuously updated model of a set of minds engaged in fragmentary symbolic exchanges of orders, questions, and information, referring to and interacting with a physical and psychological background.

Its output a particular favorite of Churchill, Turing's group obtained more and more information, and regulated the use of its own output, even successfully calling for particular military actions to check or obtain information, or to mislead their opponents. Similarly, while the functional simulation of the purely formal, modular transformation of symbols has proved to be the linguist's high road into the mind, this enterprise must eventually set these inner formal processes within the outer world that explains much of their utility and reference.

I cannot leave Turing's military excursion without noting some ironies human psychology reveals in interaction with new science and technology; they are still instructive today. Throughout Bletchley's notable successes the Germans had several major indications that their ciphered naval communications were available to the British. Yet German faith in the Enigma technology always led them to blame British spies, lurking in dockyards, preying on human frailty. Similarly, over half the German military research and development budget went to the development of the V2 rocket, a 'miracle weapon' which even when finally operational was more expensive than its limited destructiveness could justify – though it put its chief creator and propagandist, Werner von Braun, well along on his personal, lifelong goal of making a space ship.

On the other hand, if the British government's much smaller investment inadvertently advanced Alan Mathison Turing's ambition to 'build a brain', it most certainly got its war's worth.

Turing played a significant role in the first two British postwar computer construction projects. He had the satisfaction of naming the National Physical Laboratory computer Automatic Computing Engine (ACE), so recalling Babbage's Analytic *Engine* and he latterly worked with the Manchester project. He made many of the initial right moves in the direction of powerful computer languages, and he briefly played the role of prophet for the computer age, writing three powerful papers that startle and reward us still today. He lacked, however, Braun's penchant for administration, particularly when the giddy make-do rush of wartime gave way to the economies and politics of peace. A dangerous forthrightness led him to admit homosexual activity to the Manchester policemen he called in to investigate a robbery and blackmail attempt; as an alternative to prison, he was forced to take massive hormone doses that made him impotent. Though Wittgenstein died of natural causes, his last words an improbable insistence that he had led a happy life, Turing quietly dispatched himself by biting into an apple laced with cyanide.[4]

We turn first to the paper in which Turing undeniably set the terms of the debate on whether a computer can think. Latterly, we shall take a briefer look at his startlingly trenchant anticipation and analysis of what today's fashion considers the most exciting and pressing current question facing artificial intelligence work.

9
The Imitation Game

Could a machine think? – Could it be in pain? – Well, is the human body to be called such a machine? Surely it comes as close as possible to being such a machine.

But a machine surely cannot think! – Is that an empirical statement? No. We can only say of a human being and what is like one that it thinks. We also say it of dolls and no doubt of spirits too. Look at the word 'to think' as a tool.

Wittgenstein

The original question, 'Can machines think?' may be too meaningless to deserve discussion. Nevertheless I believe that at the end of the century the use of words and general educated opinion will have altered so much that one will be able to speak of machines thinking without expecting to be contradicted.

Turing

We may hope that machines will eventually compete with men in all purely intellectual fields. But which are the best to start with? Even this is a difficult decision. Many people think that a very abstract activity, like the playing of chess, would be best. It can also be maintained that it is best to provide the machine with the best sense organs that money can buy, and then teach it to understand and speak English. This process could follow the normal teaching of a child. Things would be pointed out and named, etc. Again I do not know what the right answer is, but I think both approaches should be tried.

Turing[1]

In 1950, Turing began 'Computing Machinery and Intelligence' by suggesting that neither dictionaries nor polls can give us a satisfactory answer to the question of whether computing machinery can think – or as *to what thinking is* in a way that would decide the question.

Indeed we each have mastered a narrative, intentional idiom, mastered a set of tools for characterizing – 'personating' one might call it – fellow humans and our self. Despite all the roughness, contradiction, and incoherence that Wittgenstein shows us, we are thoroughly *used to telling* what other humans think and believe, want or intend, *telling* whether they are bright or stupid, thoughtless, cruel, or incoherent, and so on. Wittgenstein recalls us to Descartes' and La Mettrie's fascination with dolls, whose simple mechanical innards, incomparably more open to clear view than our own, both charm and, as folk tale and literature show, can deeply frighten us. Of them, Wittgenstein flatly asserts *we can only say* that they think (believe, etc.) just as we so speak of humans and what we choose to find is like humans.

If we robustly deny that a machine can think (even in the face of the example of the human body), this represents a clash of diverse idioms, not a matter of fact. Further, given our easy extension of the intentional idiom to dolls and spirits, we may surely (as Turing predicted in 1950) extend the idiom to computers. Indeed, one has but to listen to human talk today to realize how far we have gone in confirming Turing's confident prediction. Recalling our central theme, the inverse adaptation can appear more salient, diverse, and enlightening: we seek *input*, we *deprogram*, we *go online*, our mind is the *software* of our *hardware* brain, though some of it is *hardwired in*; neurons are *flipflops* but though we *information process* we are *parallel* not *serial* processors, etc.

If, following Wittgenstein's suggestion, we wish to use the word 'to think' as a tool, how should we employ it?

Though we have an idiom, we do not have a clear conception of what thinking is, what cognition and intelligence are. And while we might accept the shift from consciousness to computation suggested by Wittgenstein and Turing, we do not know what style of computation, what combination of input, output, memory, and computational technique, mounts up to real intelligence, real thinking. Turing put the problem bruskly:

I propose to consider the question, 'Can Machines Think?' This should begin with definitions of the meaning of the terms 'machine' and 'think'. The definitions might be framed so as to reflect so far as possible the normal use of the words, but this attitude is dangerous. If the meaning of the words 'machine' and 'think' are to be found by examining how they are commonly used it is difficult to escape the conclusion that the meaning and the answer to the question, 'Can machines think?' is to be sought in a statistical survey such as a Gallup Poll. But this is absurd. Instead of attempting such a definition I shall replace the question by another, which is closely related to it and is expressed in relatively unambiguous words.
The new form of the problem can be described in terms of a game which we call the 'imitation game'.

Startlingly and disarmingly, Turing now describes a contest between three people, 'a man (A), a woman (B), and an interrogator (C), who may be of either sex.' In another room and so barred from using physical appearance as opposed to cognitive evidence, the interrogator, through questions, is challenged to tell which is which with the cooperation of the woman but not the man (so that 'tones of voice may not help the interrogator . . . the ideal arrangement is to have a teleprinter to communicate between the two rooms'). Turing suggests the interrogator might well begin by asking about the subject's supposedly feminine hairstyle, thus perhaps eliciting from A, 'My hair is shingled, and the longest strands are about nine inches long.' Turing mentions that B may indeed say, 'I am the woman, do not listen to him,' but he adds that this will 'avail nothing as the man can make similar remarks.'

Turing suggests that the best strategy for the women would probably be simply to tell the truth, to perform as best she can on the litmus tests the interrogator might think of, to give as honest and detailed a description of her life and thought as possible, perhaps making suggestions about what a man would be unlikely to know or be revealingly slow and clumsy about, or suggestions about questions that a male might answer too quickly or aggressively, and so on.

The male, of course, can be imagined to be racking his brains to be sure his impersonation never slips in vocabulary, turn of phrase, in self-characterization, style and speed in solving problems, arguing, narrating, etc. His impersonation is much more demanding than that of a male novelist or actor, for the novelist can take all the time he wants, expose his female character only to certain questions and problems, and then correct and polish, while the actor, though he must perform in real time, already has his lines and gestures set. Turing is surely right in presuming that if the man's impersonation succeeds, we can plausibly conclude that *he can think like a woman*.

Having explained the imitation game, Turing now proposes his question, 'What will happen if a machine takes the part of A [the man] in this game?' This is what we now call the Turing test.

For the male to be able, in the first test, to think like a woman is of course to be able to think, period, and to think with great skill. Analogously, Turing suggests that *being able to perform indistinguishably from a human thinker is to be able to think, period*. The imitation test format 'has the advantage of drawing a fairly sharp line between the physical and intellectual capacities of a man . . . We do not wish to penalize the machine for its inability to shine in beauty competitions.'

Turing notes that we should want to leave room for the possibility of machines, or Martians, who think but who, even with extensive preparation, cannot do a creditable job of impersonation. But since, lacking a clear definition of thinking, we at least seem confident that we humans are thinkers, so we surely must accept success at *this* test, anyhow, as a positive demonstration of thinking, of intelligence. Turing test passage is sufficient for intelligence, though it is unlikely to be necessary.

Artificial intelligence work has developed as a multifaceted pursuit of the task Turing proposes, both in terms of general reasoning programs and, more commonly, particulars from chess playing, which Turing had already worked on, to the recent 'expert systems' for, e.g., specialized medical diagnosis. Understandably, since much of this has immediate practical use or more narrow theoretical significance, and since such work can be pursued without deciding what Turing test passing might mean if we ever achieve it, practitioners have often not felt a need to say or even consider whether a successful passer would be a real thinker, an intelligent being, or just somehow a clever contrivance, a mechanical, unthinking imitation.

The word 'simulation' helps this ambiguity along in two ways. The first is that we use the word for models of physical events, characteristically of processes too large, expensive, or quick for direct experiment and observation. But a computer-simulated hurricane is not a hurricane (fortunately), and the pins placed to show the latest Atlantic U-boat positions are not themselves U-boats. The second is that 'simulation' like 'artificial' may sometimes suggest illegitimacy: a counterfeit 100-dollar bill is real paper and indeed may even possibly be physically indistinguishable from properly issued currency, but it is not 100 dollars. As to the first, the input to a hurricane model are numbers that represent initial weather conditions, and not of course actual winds. But the input to A and B in the Turing test is identical, both are asked the same questions, not one blasted by wind and the other by numbers; and if the same output print patterns soon appear on both screens, they count as the same sentences.

That Turing calls for an *impersonation* ingeniously meets the second charge of illegitimacy by co-opting it. We imagine a female sociolinguist and playwright saying of yet another unsuccessful candidate, 'A is not convincing: first we have a caricature with the dutiful, sugary phrases overdone – and then when A tells the story of A's supposed problems with "her" little girl, A envisions her as a father might, not a mother, [etc.]'; other team members offer equally careful analyses in this and the other cases. But now, through a bravura display, a male passes as a woman under batteries of tests and meticulous personal questioning ('My hair is . . .' etc.). Given that, even under such scrutiny, A passes, it would surely be unsporting for someone now to protest 'No, that does not show that he can think like a woman because *he is* NOT *female.*'

What, after all, is *impersonation* supposed to be? It is absurd to say that someone's impersonation of, say, Mikhail Gorbachev is, however excellent otherwise, fatally flawed unless the impersonator *actually is Gorbachev*. (I must confess that, however absurd, a few of my undergraduates have, as the Germans had for their Enigma Machine, complete and unsupported confidence that computers will soon, indeed probably *already* easily can, meet Turing's actual standard. Of these, the majority triumphantly protest that the computer, none the less, cannot really pass the test, *because* it is *not a human being*, while a minority, with equal absurdity though rather more charm, conclude that since the computer can pass the Turing test, it must *be a human being*.)

Of course, we recognize the emotion that sparks this protest. We understand only too well the baffled outrage of Victorian male critics when the master novelist, George Eliot, is unmasked as Mary Ann Evans. We expect the rereading and downgrading as much as the opposite, equally suspicious move, the search for the male who really must be responsible, the legitimate author or the mentor, father, corrector, inspirer ('the programmer'). When Mary Wollstonecraft wrote *A Vindication of the Rights of Woman*, she challenged a tradition almost unquestioned since Aristotle's assertion that woman could at best only take the imprint of male thought – mere giant, featherless parrots as several eighteenth-century male writers contended. Wollstonecraft pleads that females have much the same cognitive equipment as men and will, with equal entry to education, impress the unbiased with a similar show of cognitive acquisitions. Analogous arguments have been made for disadvantaged racial and ethnic groups, indeed for all whose physical appearance, name, birthplace, actual identity, etc., might prejudice judgement; hence we have 'blind' refereeing of written submissions to contests, academic examiners, and scholarly journals, and a variety of similar practices for hiring and advancement, for commercial products and construction projects, etc. And so, similarly, Turing called on us to 'play fair' with the machine.

In *Computer Power and Human Reason* (1978), MIT computer scientist Josef Weizenbaum leveled a jeremiad at an early move in the change in everyday talk about computers that Turing predicted. His short, very limited ELIZA program simulated an interview by a nondirective psychotherapist. ELIZA introduced itself, asked for your name which it would use thereafter, responded to your statements about yourself with a few all-purpose commentary sentences into which it would fit key words such as *father, boyfriend*, etc., if you had typed them in your most recent input. Weizenbaum was shocked that some people, who were unfamiliar with computing or the simple tricks of programming, would slip into the intentional idiom when talking about ELIZA, thus in small ways speaking of it as a person. Of course, even in this restricted format ELIZA often produces nonsense, and almost any direct question will expose it.

More recently, the much more complex and versatile PARRY program simulates a paranoid patient well enough so that psychiatrists are unable to distinguish it from a human paranoid *within the format of a psychiatric interview*. Still, the result is very fragile indeed. Not only could the psychiatrist have resolved his doubts with a series of real world questions and tests of common knowledge, but the human paranoids were *not* told what was going on and so they were not encouraged to say things that might have facilitated the psychiatrist's decision.

But these two programs are minor anecdotes that emphasize, along with failure and limited progress with larger-scale problem-solver and PROLOG projects and narrower expert systems, how far we are from producing a serious contender, and the predictions issued from MIT and Pittsburgh in the 1950s that talked in terms of five and ten years are now more guarded, though today's Von Neumann Machines offer processing speed and memory capacity far beyond what Turing and others imagined would be required. Today, the impressively financed, large-scale and fairly long-range Japanese 'Fifth Generation' project modestly aims, we are told, to achieve the intelligence of a normal human 5-year-old. This has about it the misleading suggestion that we have done the 4-year-old or maybe just the 3-year-old but we have done nothing of the sort; moreover, since now most cognitive scientists would agree that *if* we could construct a reasonably principled rendition of a 5-year-old, the 12- or 20-year-old would be an unproblematic extension, perhaps achieved simply by letting our 5-year-old learn for a few years, so there is really nothing in the least modest about the announced goal.

It is quite clear that we are still nowhere near producing a Turing test passer or near passer. Still, instructive and sometimes extraordinary progress has been made, along with equally instructive failure, and the progress continues in a zigzag fashion. All the while, as spin off, we have made a startling variety of theoretical and practical discoveries in programming and hardware. Above all, this complicated boot-strapping enterprise has taught us (as we have tried to make various aspects of our intelligence and knowledge clear and explicit, particularly by the stern standards imposed by the electronic digital computer and its programming languages like LISP), that we actually understand shockingly little about how we come to understand, and deal intelligently with, the world around us.

Forgoing the naive, Germanic supposition that the all-powerful machine will easily pass Turing's test and so the test itself must be ruled out in principle, we are gratified to realize that our cognition, particularly what is rooted in our common biological endowment, is much richer, much more complex and powerful, than it had seemed. The price of this gratification, however, has been the realization that the richness we have discovered largely operates *outside* of our consciousness: dazzlingly complex computational processes achieve our visual and linguistic understanding, but apart from a few levels of representation these are as little open to our conscious view as the multitudinous rhythm of blood flow through the countless vessels of our brain.

In 'Computing Machinery and Intelligence', Turing succinctly replies as well as anyone could today to objections to the very idea that a computing machine can think.

To 'arguments' that a machine cannot 'be kind, beautiful, have initiative, fall in love, enjoy strawberries, make someone fall in love with it,' etc., he replies devastatingly,

The inability to enjoy strawberries and cream may have struck the reader as frivolous. Possibly a machine might be made to enjoy this delicious dish, but any attempt to do so would be idiotic. What is important about this disability is that it contributes to some of the other disabilities, e.g., to the difficulty of the same kind of friendliness occurring between man and machine as between white man and white man, or black man and black man.

Characteristically, Turing treats 'Lady Lovelace's Objection' respectfully and ingenuously, absolving her 1842 remark that 'The Analytic Engine has no pretensions to *originate* anything.' None the less, while computers as much as humans may have much built into them and learn much through instruction, their behavior can be startling and utterly unexpected. One of the first theorem-proving programs came up with a new shortcut proof that completely surprised its programmer. The recent University of Illinois computer proof that four colors suffice to partition a map, thus solving Isaac Newton's puzzle, was not only a first but the proof is so long that no human mathematician can check it, a marvelous refutation of what Turing dubs the fallacy of philosophers and mathematicians, the Cartesian presumption:

that as soon as fact is presented to a mind all the consequences of that fact spring into the mind simultaneously with it. It is a very useful assumption under many circumstances, but one too easily forgets that it is false. A natural consequence of doing so is that one then assumes that there is no virtue in the mere working out of consequences from data and general principles.

Turing reminds us that a theorem or a sentence has form and meaning only within a system of implications and that what is original or surprising may be, perhaps always is, the conscious or unconscious 'working out of consequences' from theory and data.

The illustration of this point, in interchanges between Dr Watson and Sherlock Holmes, is a basic feature of Conan Doyle's stories. In nearly every story, the great detective astounds Watson by announcing, after inspecting the crime scene, that the killer was 'more than six feet tall, had small feet for his height, a florid face, long fingernails, and smoked a Trichinopoly cigar', or after inspecting a pocket watch, concludes that the owner is a once well-off alcoholic with money problems, whose wife has recently ceased to love him, and so on. However, when Holmes explains how he reached these conclusions, Watson invariably says that he had thought that Holmes had done something truly impressive but now realizes that it is all commonplace. This reaction invariably enrages Holmes. He will lecture Watson to the effect that all real cognition, whether everyday or unusually complex and original, is data collection, analysis, and deduction, though some steps may be automatic or uncon-scious – a mere random guess counts for nothing, though it will, like a lottery ticket, occasionally win. The most exceptional and ubiquitous feature of Holmes' performances is *speed*, the rapid and relentless crime scene data collection accompanied by the rapid initial deductions that impress Watson, and then the single-minded search, lightning deductive chain after chain after chain that may cease only after Holmes has consumed the hours of the night and several pipefuls of tobacco, having now searched through all the possible configurations that might have generated the evidence. Holmes will finally end his lecture by threatening never to tell Watson how he deduces his initial oracular assertions, so that he will reap and keep Watson's astonished awe, even though he gains it for the wrong reasons. Fortunately for us, Holmes' temperament makes it impossible for him to execute his threat.

We do well to remember Holmes' lecture and Watson's bumbling example of an all too human, self-serving fallacy, one that fends off Mettrie's characterization of the soul as the organization of an enlightened machine. Turing reminds us, through the metaphor of peeling off the skins of an onion, that our investigations of human cognition may exhibit a Watsonian quality,

In considering the functions of the mind or the brain we find certain operations which we can explain in purely mechanical terms. This we say does not correspond to the real mind: it is a sort of skin which we must strip off if we are to find the real mind.

Though Turing expects that we shall 'eventually come to the skin that has nothing in it', he recognizes the power of the presumption that whenever we achieve a real, step by step understanding of some part of our cognition, we tend like Watson to discount it as 'mere working out', not 'real' intelligence at all. What the onion skin metaphor can also suggest is that intelligence is a characteristic of the whole system of layers. If we none the less insist that no matter how many layers have been understood, there must be a 'creative inner act', a mysterious supernatural core known only through individual experience, we have retreated into solipsism.

Turing gives shorter shrift than Wittgenstein to the straightforward 'solipsist' objection.

According to the most extreme form of this view the only way by which one could be sure that a machine thinks is to *be* the machine and to feel oneself thinking. One could then describe these feelings to the world, but of course no one would be justified in taking any notice. Likewise according to this view the only way to know that a *man* thinks is to be that particular man. It may be the most logical view to hold but it makes communication of ideas difficult. A is liable to believe 'A thinks but B does not' whilst B believes 'B thinks but A does not.' Instead of arguing continually over this point it is usual to have the polite convention that everyone thinks.

The *reason* that you and I have for thinking that other human beings think is that *they pass the Turing test*. Of course, we assume that something that looks human and healthy and awake can think because the two often go together, but proof positive, both psychological and legal, requires *and requires no more than* linguistic performance, some approximation of Descartes' 'reply appropriately to whatever is said in its presence'.

In literal fact, Turing's 1950 paper, in proposing the imitation game as a test, was addressed to human readers. Turing was rallying the humans, not the digital electronic computers. He challenges us to look clearly at our cognitive qualities, *to show we can know how we work, cognitively, to show we can try to be Holmes, when possible, and not Watson*, and get down to this most Socratic of tasks, self-knowledge. The Turing test is our test, the how-to-know ourselves test.

How to pass it?

Turing, who operationalized the question *Can a machine think?* into *Can we make a Turing test passer?*, now suggests that we operationalize rationalism and empiricism as two over-arching techniques for making a passer, 'both of which should be tried.'

Turing, as usual, allows for possibility of complete physical duplication. If, as Dr Frankenstein, we are able to replace *every* part of a human, brain to buttocks, legs and toes, warts and all, and then send the result out to learn the ways of the world and wait for a few years, then we shall obviously have a cognitive replacement, a Turing test passer, but we have duplicated everything else as well. Turing, surely rightly, suggests that much of this, like skin color and slant of brow, cannot be relevant to a Turing test pass. What may not be irrelevant is a span of experience (learning, training) by means of 'the best sense organs that money can buy' which will make a thinker out of an 'unorganized machine'. Momentarily embracing this empiricist technique, Turing warms to this operationalized version of John Locke's blank tablet,

Presumably the child brain is something like a notebook as one buys it from the stationers. Rather little mechanism, and lots of blank sheets. (Mechanism and writing are from our point of view almost synonymous.) Our hope is that there is so little mechanism in the child brain that something like it can be easily programmed.

However, most unfortunately for the Socratic quest,

An important feature of a learning machine is that its teacher will often be very largely ignorant of quite what is going on inside, although he may still be able to some extent to predict his pupil's behaviour. This should apply most strongly to the later education of a machine arising from a child machine of well-tried design (or program). (p. 34)

You see what Turing has done. The traditional empiricist understood intelligence as arbitrary, environmentally induced, associations established by experience and maturation between *our subjective, conscious experience of* sensory inputs and motor outputs. The twentieth-century behaviorist, similarly, understands cognition as the arbitrary structure of regularities in behavior correlated with positive and negative reinforcement from the environment. Neither empiricist thesis openly commits itself to experimental contest, nor to a full, procedural, and mechanical account. Turing suggests just such a commitment.

The testable, technological claim is that we shall find the 'unorganized machine' plus 'training' the most successful way of simulating human intelligence. Turing stresses that an 'unorganized machine' is not a compact, practical version of the Universal Turing Machine: if it were, then one could program it to become whatever thinker you wanted. Turing imagines something like a network of connections which will alter and develop through hundreds of thousands of environmental exposures, gradually acquiring organizations built upon organizations of the connections between inner units analogous to nerve cells. In 'Intelligent Machinery', Turing describes 'A-type unorganized machines' which 'are of interest as being about the simplest model of a nervous system with a random arrangement of neurons'. Like neurons, the machine's nodes are connected by input or output lines to some of the other nodes; at each transitional 'moment' a node will or will not 'fire' depending on the input weight and the node's own state. Such machines settle into 'periodic' behavior after a finite number of moments and 'machines of this character can behave in a very complicated manner when the number of [units] is large'. Through a long period of 'training' the properties of the network may be altered into an intelligent configuration. Turing also mentions a 'B-type unorganized machine' which is a network in which some or all of the nodes are themselves computers, thus able to perform what is today called 'massively parallel distributed processing'.

Indeed, placing an emphasis on rationalist native endowment, Turing imagines that much design and experiment will be needed to determine the appropriate initial state and architecture of the child machine; he suggests we may hope not to need as much of this design and experimentation as nature needed to evolve the human. The *more* cognition that needs to be built into the child machine, the less will be determined by the 'training' process. Eventually perhaps, 'the best eyes and ears that money can buy' will help structure the appropriate initial machine well enough to produce adequate visual and verbal perception and, eventually perhaps, natural language understanding. This last will expedite learning and make the child machine into some approximation of a Universal Machine, and of course it is a prerequisite for Turing test passing. Increasingly, over the past decade, many cognitive scientists have called for a switch to this approach (though they have often ignored the possibility that the child machine might need to be richly structured in its initial state). They expand on Turing's argument from the analogy between nerves and networks, to the analogy between human development and 'machine-training'. While this approach was briefly touted by Frank Rosenblatt in the 1960s, and then formally examined by Minsky and Seymour Papert in *Perceptrons* (1969), perhaps the best reason for the recent enthusiasm for this approach is the unexpectedly slow and fragmentary progress of a 'naive' rationalist program, particularly the one that easily assumes that a programming language like LISP provides the 'initial state' and 'architecture' in which to encode the data and rules.

Turing also operationalizes rationalism as success in program writing, starting perhaps with more formal problems like chess playing, theorem-proving, decoding, language translation, problem-solving, etc. Ultimately, the aim is to build a suitably powerful computer and program it with a data store and structured set of rules that will allow it to pass the Turing test.

Turing himself, naturally, prefers the rationalist program in that it holds promise of a more thorough understanding. Setting the pattern for many AI enthusiasts, he was prone to assume that he himself understood, or could easily call to mind, most if not all of his 'cognitive routines' so that all he, and his co-workers, might need to do is to order the machine to act in much the same way. Since human brains may be capable of storing not a great deal more, Turing suggests a machine memory capacity of one hundred million bits (which he airily describes as equivalent to the capacity of the *Encyclopaedia Britannica*, 11th edition). Since this was (and now is) well within reach, the real problem would be writing program. 'At my present rate of working I produce about a thousand digits of program a day, so that sixty workers, working steadily through the next fifty years might accomplish the job, if nothing went into the waste basket'. No wonder he wryly comments, 'Some more expeditious method seems desirable.'[2]

He suggests we 'first put a program into it which corresponds to building in a logical system (like Russell's *Principia Mathematica*)': to do this is to adopt a 'higher level programming language'. After adding, *in that language*, various general principles and directives, one might need only to type in a pedestrian rendition of the facts, attitudes, and know-how normal to, say, an English-speaking, literate, well-informed 1990s adult human, plus the specialized mix of these and specific invented biographical data, that would inform Turing test passage.

While Turing himself did not part company with his goal, he does mention that the majority of the cerebral cortex seems dedicated to visual processing and visual memory and he seems rather to hope that Turing test passage might occur without having to simulate much of this. If we, in addition, supposed that the 'specialized mix of facts, attitudes, and know-how' and a 'biography' might be a trivial, though tedious, matter to fill in, and that even the more general store of a 1990s adult would present no basic problems, then we might conclude that at bottom intelligence could really amount to no more than a sophisticated programming language and a problem-solving executive. Subtract, so to speak, all particular data and all particular knowledge – all that some humans have and others do not – and then you have the essence of intelligence.

Subtract all particular data and knowledge and what is left is intelligence is the easy sort of equation that Turing's notion of Turing Machine and Universal Machine warns us against. Recall the simplicity of the TM, got by farming out everything possible to the tape. An 'adder TM' might be first fed some instruction, some data, then print on tape to 'carry', then another instruction, a 'carry' read, and so on, until the long tape/machine interaction ends with a final print out. The starkly minimal

UTM has no logical circuits, no Von Neumann stock of hard-wired operations, much less any sort of higher-level programming language or general knowledge: there is no absolute distinction between intelligence and knowledge to be had here, nor between program and data. Indeed, the point is that in what Turing calls 'practical computing machines' (ourselves a prime example) these tradeoffs are determined by practicable design possibilities, by what can actually be constructed, given constraints of speed, reliability, materials, etc. (whether the constructor is biological evolution or the cognitive scientist/engineer).

Above all Turing wanted to say that the Turing test was a real test, an extraordinarily challenging one: an invitation, or opportunity to vie, to use the oldest sense of the word. A *real* test meant one it was most necessary that we actually work out the various approaches in detail, taking on the full burden of the storage and processing requirements, the weave of developmental, explicit programming and training/learning routines: this is the most expeditious way to also understand important aspects of our own cognition, for we shall have been cut by the same, or related, considerations. We may for example, Turing suggested, hope to arrive at the appropriate 'child machine' without as protracted a design period as nature used, but he did not take this for granted and he very certainly supposed that we should expect simulations of our cognitive development and our intelligence to be constrained by considerations of design and processing that are usefully compared. Nature will seem to imitate art as well as the other way around. The translation function that translates envisionable computerese into natural language can be worked on from either end.

In suggesting that we might start with coding/decoding and machine translation (chess playing, mathematics, etc.) *or* with building a 'child machine' that will be sent to school where much instruction will go on in a human language, Turing recognizes the central importance of natural language. The Turing test itself insists on this centrality, for the questions and the replies are all sentences in English (or another human language). Accordingly, we shall turn to human language in the next chapter and to the human 'child machine'.

I shall end this chapter by considering two recent mistaken attacks on Turing's project. Both seem to involve the Watsonian fallacy and, implicitly, fail to take seriously the step from the conscious to the computational mind, from inner theatre to thinking machine, from a black box with a dream inside to functional and practical models of real cognitive mechanism. They misunderstand and mistake Turing's invitation to cognitive science.

By a Turing Machine we can mean some particular literal version of the minimalist machine that Turing described, with its very long tape of 1s and 0s, *or* we can mean any *mechanical device* whose innards, through whatever system of shortcuts, achieve the same computational properties (I emphasize *mechanical* because a 'black box' which temporarily produced the appropriate outputs through supernatural miracle or a lucky string of random accidents would not be a Turing Machine in either the first or second sense). Current cognitive science emphasizes the second sense in calling itself *functionalist*. This has also led us to speak of each particular human being, cognitively speaking, as a particular Turing Machine. Speaking at a much greater level of generality, it also leads us to say that we, and our digital electronic computers, are Universal Turing Machines.

Though he puts the point jokingly, Turing excludes the most economical way to produce a Turing test passer – biologically conceive and bear a human child in the usual way and help it to develop into a normal adult. The challenge of course is to *build*, to design, construct, program, and train, a Turing test passer, and do so to understand more clearly how we meet the same requirements. In 'Troubles With Functionalism' Professor Ned Block imagines that we hire out the adult population of China to enact the literal Turing Machine, with its minimalist architecture and endless instructional tape, that has the same cognitive properties as some actual human being, call him Adam: the Chinese assemblage would give the same answers to questions, display the same cognitive strengths and weaknesses, etc. Here, it would seem, we have a Turing test passer *and*, and this is Block's point, we are *not* inclined to believe that this massive assemblage thinks and thinks like Adam, etc. So, the Turing test is no good because here you have a passer that is not a thinker.

Turing's simplest response would be to point out that the collective Adam would not in fact be a test passer. If you recall the relentless way in which Turing's minimalism converts complex two-dimensional input into miles of 1s and 0s, changes alphabets, ten-base numbers, and simple arithmetical operations into more miles, etc., you realize that the Chinese assemblage might take months to grind out responses that the human Adam would produce in seconds. Moreover, disease, human frailty, and the endlessly repetitive nature of the ballet will surely make it unaccountably unreliable (they would be better off doing a literal imitation of a Von Neumann machine, a 'practical computing machine' as Turing labeled it; but the growing suspicion that even the fastest and most reliable Von Neumann machines have an architecture that is grossly inefficient for simulating human cognition makes one fear that the Chinese assemblage would not come anywhere close even with this switch).

Professor Block might respond to the reliability problem by suggesting we 'just suppose' that this one particular time it all happened to work out right. But this is really just the notion of supernatural miracle or wildly improbable luck: the assemblage is not a *machine*, that is, a physical device whose output is reliably and mechanically related to the physical properties of its innards, whose nature is to be identified with the properties of this inner system.

What is of course true is that there is a Turing Machine that specifies Adam's cognitive properties, but we know this simply because we see how complicated and compact cognitive properties can be broken down into simpler but bulkier ones. Similarly, you will recall, the actual, printed *PM* specifies its own conversion into an almost unprintably long version, which, when literally described in unabbreviated Gödel numbers, becomes in practical physical terms unprintably long. Familiarly, the difference between the two cases is that we understand *PM* and its properties in fairly clear detail (that is why building *PM*, or some similar logical language, into the computer has seemed such an obvious first step). On the other hand, we do not at all have a clear formal account of much of the rest of our cognition. It is elementary that, *once having built the actual book*, so to speak, that inscribes Adam's cognitive properties, we shall *then*, trivially enough, have determined the literal Adamite Turing Machine and determined it from the only practicable direction. We have reason to believe that this literal Turing Machine effectively could not be built, given the limits physics imposes on mechanisms.

We could improve the design by replacing the Chinese with the simple mechanical parts that they are simulating; we could try to build *the literal Turing Machine* that has Adam's cognitive properties from reliable parts. You will notice that this does not solve the time problem – nor even, entirely, the reliability problem when one thinks of thousands of miles of tape. So, seemingly paradoxically, we realize that the literal Turing Machine that exemplifies Adam's cognitive properties cannot literally pass the Turing test, cannot reliably and in real time simulate Adam's intelligence, though it is of course a theoretical description of Adam's intelligence and of a properly constructed and programmed computing device that compresses and shortcuts its way to success in simulating Adam's cognition. Turing often called Turing Machines 'theoretical computing devices' as opposed to the 'practical computing machines', the digital electronic computers, that he helped design and build.

In *Behavioral and Brain Science* and, mostly recently, *Scientific American*, Professor John Searle imagines that he is boxed. Chinese sentences are sent in on slips of paper, Searle consults a 'Chinese Turing Test Crib Book' to find the appropriate output, and then he sends out the listed response. Searle does not understand Chinese at all. He simply finds the symbol sequence on the left-hand column of his crib book that matches the input; then he issues the symbol sequence he finds in the right-hand column on the corresponding line. While Searle starts off from AI attempts to simulate common human knowledge of what goes on in having a meal in a restaurant or taking a commercial air flight, his general target is the Turing test. At the end of this process, Searle argues, he surely has not *passed the Turing test in Chinese*: he does not understand or mean the Chinese sentences – and still less is this true of the slips of paper and the crib book. As contrast, Searle insists that if he had performed equivalently with English, his native tongue, then there would be genuine understanding and meaning. But if there is no meaning or thinking going on in the Chinese box, then, Searle concludes, we should be foolhardy indeed to think that if a formal symbol-crunching electronic computer 'passed the Turing test', it *must* be a thinker, for the computer would be doing nothing different from what Searle was doing, namely, symbol shuffling.

Searle thumbs his nose at time and reliability issues with Watsonian abandon, for Chinese, given its several thousand nonalphabetical ideo-grammes, is just the human language to make his procedure most clearly utterly impractical from the first step. He might well need hours to look up his first symbol and days, if not weeks, to make a single response. While a computer undoubtedly could carry out Searle's procedure much more quickly, the procedure consumes time so prodigally that it has been the first, among cognitive scientists, to be ruled out as practically impossible for the fastest electronic computers *and*, quite independently but in striking parallel, for the human neurological computer.

Rather more floridly than Block, Searle suggests in effect that the computing machinery that passes the Turing test, that regularly achieves this in mechanically and programmatically explicable ways, is JUST a fancier, or merely quantitatively augmented, version of Searle shuffling slips of paper at about the level that might produce a reasonable perform-ance at tick-tack-toe. And it *is* a charming achievement to make a (reasonably) successful tick-tack-toe player out of wood tinker toys: the wooden joints may stick or push out of designed true, and a little too much is being asked of structure and material, but the device, mostly, works and it is clear, and revealing, why it works, and it is perfectly legitimate to call it functionally equivalent to the structures employed by electronic computers and humans.

However, it explains nothing to say that the tick-tack-toe tinkertoy is virtually a Hi-Tech chess player or what have you, even a Turing test passer and hence a serviceable good thinker (or *mean*er), though a little rough or not fully worked out – and then following this with the complaint that the awkward movements of wooden rods and wheels do not seem to be thought or meant, do not seem adventuresome or tricky, forced strategically or short-sighted, etc.

Descartes, long before Turing, noted that·we can easily build clockwork gadgets that might say 'I am in pain' when touched (or, one might add, say, 'I think therefore I am' at random intervals). But this would in no way be to demonstrate intelligence, in no way be to 'reply appropriately to whatever is said in its presence' and so pass Descartes' 'one sure test'. Noises issued are meant and understood sentences only within the context of a whole productive and interpretive set of mechanisms.

Turing wrote that the bare question *can a machine think?* can seem answerlessly vague. Like Wittgenstein, he saw the human being as the one clear example of a thinking machine, and he would have agreed with Wittgenstein that looking at a human as a person or as a machine may well amount to no more than a duck-rabbit choice of attitude (the *details* that assemble into an attitude may be, however, testable and most compellingly revealing). This question of attitude has all the lack of empirical answer that Wittgenstein and Turing suggest: and it is knit with the solipsistic indeterminacy of mind as inner subjective experience, with the compelling sense that my words, as *I* think or say them, glow with real meaning, picture facts, point to objects, etc., with my magical sense that I know myself to be a thinker in each and every moment of consciousness. While doubtless Turing is right in predicting that we humans shall continue increasingly to speak of computing machines in the intentional idiom, this shift itself proves little. On the other hand, the challenge to build a Turing test passing computing machine – Turing's substitute question – can produce much more definite and specific results, and more beautiful, exciting, and useful ones as well. Moreover, *if* we do manage to produce fully convincing simulations of human intelligence, produce machines that can 'pass' as human, indeed a bright and ingenious human, we shall inevitably and often insightfully both talk about computers as thinkers and persons *and* also find ourselves happy seeing humans, with revealing specificity, as machines. It is the empirical investigation, the creative fabrication, that will carry the day because it will fill it with discovered hard detail and richer theoretical outline.

In a passage that gives such fresh specificity to Wittgenstein's comments on seeing people as minds or machines, or seeing marks as meaningful or meaningless, Douglas Hofstadter notes,

If someone were to write a program that could deal in Chinese with simple questions and answers about restaurant visits, and if that program were in turn written in another language – say, the hypothetical language 'SEARLE' (for

'Simulated East-Asian Restaurant-Lingo Expert'), I could choose to view the system either as genuinely speaking Chinese (assuming it gave a creditable and not too slow performance), or as genuinely speaking SEARLE. I can shift my point of view at will. . . . If to me, Chinese is a mere bunch of 'squiggles and squoggles', I may opt for the SEARLE viewpoint; if on the other hand, SEARLE is a mere bunch of confusing technical gibberish, I will probably opt for the Chinese viewpoint.[3]

When Turing, in the 1940s, made 'paper machines' to play chess his actions bore only a superficial resemblance to Searle's imagined procedure: he knew he had taken but the first faltering steps in an eventually burgeoning and instructive enterprise, while Searle merely dismissively imagines, without bothering with a single actual faltering step, a few such steps in order to ridicule and rule out in general the most impressive and inviting response to the Socratic imperative since Socrates first said *know thyself*.

The chess world consists of 64 squares; each side starts with eight pawns, two knights, two bishops, two rooks, a queen and a king; white and black alternate in moving until one can checkmate (unanswerably threaten the other's king) or both agree to a draw. Chess notation completely describes individual moves and games. This simplicity, plus the deep complexity that has led many bright individuals to devote most of their waking hours to the game, naturally attracted the attention of cognitive scientists. As Shannon and Turing pointed out in their seminal articles on the game, a tick-tack-toe methodology *cannot* work. White has a choice of 20 moves in the opening position and the number soon rises to 30 odd; in consequence, the number of positions that can arise by white's third move is seven million, which rises to 200 million with black's reply, and estimates of the number of distinct games that can arise converge on Shannon's 10^{120} (for comparison, a mere 10^{75} seconds have elapsed since the beginning of our universe). IBM's 1990 Deep Thought, a pendulum swing toward emphasizing brute speed over program sophistication, currently examines a million positions a second and is hoped to upgrade to a billion in two years. At this higher speed, if we ran Deep Thought continuously from the beginning of the universe we should only need to multiply this 20 billion year processing time by 10^{26} to consider all the positions that chess allows. (Mathematically speaking, there is of course a Turing Machine that maps Deep Thought's processing activity through this period in a much much more lengthy form.)

Since even Deep Thought can only look forward a few moves, with no checkmates in sight, you need an evaluation metric, a comparative measure of what a position is worth, made by estimating the strength of your pieces, the space they control, the opportunities they have. These, for example, are common overall valuations: queen = 9, rook = 5, bishop = 3, knight = 2, pawn = 1 (however, center pawns usually count more than wing ones, and centralization is good for knights, but bad for kings until the end game, etc., etc.); but all estimates are provisional – a lowly pawn, for example, can count as 18 if placed to capture a queen on the eighth rank. With an evaluation system, the computer can decide which positions are worth pursuing and 'prune' the tree of future move sequences of branches that are not worth looking at. The evaluation system represents general knowledge of chess expressed in procedural rules.

Turing, himself a mediocre player, drew from a considerable evaluative literature written for human players in producing the first sophisticated evaluation system in 1953. Expert human players actually seem to acquire a different way of seeing configurations of pieces on the board; they come to experience sophisticated estimates of the possibilities of a position in the very look of the board. And while expert players can tell you a great deal about chess, about various opening lines and combinations, they can offer disappointingly little information about how, for example, they actually put together 'the look of the board' (the expert sees *chess* while the novice sees *squiggles and squoggles*; but you cannot switch back and forth as you can with the duck-rabbit, and it takes years of intensive training to *see chess*). A marked increase in programming sophistication accounts for the leap in playing strength recently achieved in microcomputer programs like *Sargon III*. Hi-Tech, a mainframe program touted for its sophistication, recently achieved some of the first victories of computers against international grandmasters, though world champion Gary Kasparov has beaten it and Deep Thought convincingly. Deep Thought, with its emphasis on sheer speed, has also been heralded as a return to the confident early days of AI, to brute force empiricism as opposed to rationalist sophistication. In all this we have learned much about chess cognition, about human and computer players. It is extraordinary that four decades of work have produced computing machines that can vie with very bright men who have committed their lives at an early age to the deepest and most challenging of cognitive competitions. It is equally extraordinary that the best human players can still defeat the collective programmatic accumulations winnowed by these decades plus IBM's one-million-positions-a-second speed.

Speaking of 'very bright *men*' and the imitation game, chess has been the one intellectual arena in which it seemed clear that no women had or could attain the first rank; indeed, the stark standard – you by yourself win or you do not, and aesthetics and chauvinism and 'old boy' networks have little room for maneuver – that attracted cognitive scientists to chess as a test case, has made this both clear cut and troubling. Two years ago, I could have written that among the few hundred grandmasters with ratings of 2550 and higher, there are and have been no women (similarly, in tennis there are separate women's tournaments and the best woman player would probably lose to the lowest internationally ranked male player). We now have been, unexpectedly, graced by the Polgar sisters of Hungary, who play only in tournaments open to men and rather make a point of beating them. The *New York Times* describes the youngest, Judith Polgar, who in 1989 when 12 years old was rated 2550, as one of 'the four supreme chess prodigies in history' and notes that she has attained international and grandmaster levels younger than Bobby Fischer and Gary Kasparov. Here is, finally, an unequivocally great woman player, one who might well become the strongest human player in the world. When the trillion-position-evaluated-a-second Deep Thought (or the sophisticated third generation Hi-Tech) goes on line against the world's champion, it may face her.

In Ian McEwan's BBC play, *The Imitation Game* (1980), protagonist Cathy Raine, hoping to do her bit despite the granite assurance of her father, boyfriend, and society that she is a decorative nothingness, eventually becomes one of the hundreds of Bletchley women who, in complete and segregated ignorance, endlessly recorded and transcribed the radio signals that fed the fledgeling brains that Turing and other males directed. Not content with being a few peripheral flipflops, she climatically seduces a 'John Turner' into explaining the whole operation: as an outrageous, unnatural golem, 'the woman who *knows*', she must be locked up incommunicado for the duration. What most identifies John Turner with Turing is Turner's enthusiasm for 'building a brain' that would succeed in the imitation game as a thinker. Indeed a good portion of his lines are word for word from 'Computing Machinery and Intelligence'.

McEwan, known for relentlessly savaging intellectual theorists, surely began writing the play primed to detonate an ironic contrast between the unquestioned assumption that women cannot really think and Turing's blithe proselytizing for the claims of digital computers. Yet, in the final draft this particular irony was puzzlingly muted, unaccountably unexplored, and the Turingesque character, who had been Turing by name at first writing, came to be more the incidental fumbler than the antagonist. McEwan makes Turner impotent in his encounter with Raine (stealing a march on the Manchester assizes), but his flustered failure is presented as inexperience and overexcitement, and there is no suggestion that he is homosexual; though the characterization is brief and Turner not meant to shine, or burn for that matter, he is none the less the only male who is presented as superficially willing to recognize a woman as a thinking person, and he is the one male whom Raine hurts and exploits, though the drama slides by this; yet he is also overheard, in conversation, saying a sentence or two of Turing's, enough to get the idea of the imitation game as a project for proving machine intelligence. The effect is curious dramatically as if the real Turing, in his lines, had so struck McEwan that he could not let go of him, but when he began to put his imitation together, he could not make him play the role the larger drama required, and so must replace him with a diminished, dulled, uncertain caricature, like a version of *Paradise Lost* in which a minor, inexperienced sinner named Satin occasionally in passing conversation thunders out a line or two from the great defiant paradoxical speeches Milton wrote Satan.

What stayed and redirected McEwan's imitation may also have been the realization, through Hodge's research, that Turing himself suffered something akin to Cathy Raine's invalidation. And Hodge's history, giving the dramatist the boot, specifies that Bletchley held a female Cambridge mathematician, Joan Clarke, who was well aware of what was going on (just the sort that would have been locked up for the duration had McEwan been absolutely right). Clarke functioned well with Turing and the others in Hut Eight; indeed, Turing eventually proposed marriage, and the two retained a friendly relationship long after she withdrew from the engagement.

Turing hoped, indeed gaily expected, that the advent of computers would, fortunately, puncture the pretensions of an elite who could prove theorems, decipher ciphers, do mathematical calculations, parse Latin sentences, play chess, etc.; and thus strengthen the claims of humans with other, less pretentious qualities. Alan Turing correctly insisted, in the 1940s, that the first appreciable triumphs of computers would come in rapid mathematical calculation, theorem-proving, and chess playing, thus correctly reversing the predictions of male science fiction writers that persisted in denial right up to actual employment of computers in astrogation and the like. More substantially, he, in seeing clearly, set the agenda for artificial intelligence work to the most recent debates. Yet, after his death in 1954, his three most important papers were almost unavailable and others unpublished, his life and works literally unknown, buried under secrecy and prejudice, all that remained were references to 'turing machines' in computability theory and to that and the Turing test among a substantial, recently growing, number of cognitive scientists (he *was* made a Fellow of the Royal Society for his 1937 paper; his sponsor was Bertrand Russell).

One likes to think that he might have been amused that his decade-old final accession to scientific authority and acknowledgement, and especially the emblematic notoriety of recent years, when two plays about him ran at the same time on Broadway and in the West End, should have also resulted from the unashamed but hardly blatant homosexuality that cut down his access to the computers that he had, as much as anyone, created, and which also became the cause of his emasculation, perhaps his death, and certainly contributed much to the obscurity that veiled him for decades after his death.

After he finished laughing, though, I imagine Turing to add that his sexuality had nothing whatsoever to do with the truth or value of what he wrote and built. Then, surely, he would add that this equally applies to what computers might write or build. Andrew Hodges, echoing contemporary reports, writes

There was his voice, liable to stall in mid-sentence with a tense, high-pitched 'Ah-ah-ah-ah-ah' while he fished, his brain almost visibly labouring away, for the right expression, meanwhile preventing interruption. The word, when it came, might be an unexpected one, a homely analogy, slang expression, pun or wild scheme or rude suggestion accompanied with his machine-like laugh; bold but not with the coarseness of one who had seen it all and been disillusioned, but with the sharpness of one seeing it through strangely fresh eyes.

Lamenting that the BBC did not preserve its recordings of Turing's several, early 1950s radio talks on humans and machines, Douglas Hofstadter comments, 'It seems poignant to think that the voice of such a recent figure is forever lost, and all we have to go on is the written word.'[4] Turing was a long-distance runner and, barring a hip pull, he probably would have represented Britain in the 1948 Olympic marathon. Alan Turing would be 76 today. His progeny are all around us. Had he had those 25 odd years, I find it impossible not to believe that they would have been better for his schooling and building, and we better at understanding them and ourselves.

10
The Linguistic Turn:
The Child Program

The field of cryptography will perhaps be the most rewarding. There is a remarkably close parallel between the problems of the physicist and those of the cryptographer. The system on which a message is enciphered corresponds to the laws of the universe, the intercepted messages to the evidence available, the keys for a day or a message to important constants which have to be determined...

In the process of trying to imitate an adult human mind we are bound to think a good deal about the process which has brought it to the state that it is in. We may notice three components.

(a) The initial state of the mind, say at birth.
(b) The education to which it has been subjected.
(c) Other experience to which it has been subjected.

We have thus divided our problem into two parts, the child programme and the education process. We cannot expect to find a good child machine at the first attempt. One can then try another and see if it is better or worse. There is an obvious connection between this process and evolution, by the identifications

> *Structure of the child machine = hereditary material*
> *Changes of the child machine = mutations*
> *Natural selection = experimenter's judgments*

Turing[1]

We come to natural language so young and carry on in it so much of the time that we fail to notice what Wittgenstein shows us: that commonsense accounts of language learning, and of meaning and rule following, are full of explanatory gaps cloaked by myth-breeding metaphors. Further, we fail to notice the paradoxically striking difference in *what* we *see* when we see handwritten English right side up or upside down, or *what* we *hear* when we hear letters, words, and sentences of a familiar language or one that *for us* is just the auditory equivalent of squiggles and squoggles. What is it about natural language that getting one, rather than another, can so fundamentally and permanently contort and rearrange and transform what we literally see and hear? So fundamentally and permanently, indeed, that the linguist calls this 'other experience' *acquisition* to distinguish such a growth-like biological transformation from ordinary learning.

Let me mix and double Turing's metaphors as today's linguistic scientist might. 'The system of encipherment' is the human child machine's *language acquisition device*, the 'day-setting' specifies the small number of *parameters* that determine which human language you actually have, among a discrete and relatively small number that the language acquisition organ permits. (To make full use of Turing's awkward 'day or message' Bletchleyian disjunction, we could take 'message-setting' to be the way in which the onset of a variety of our native language sets off an enormously complex cognitive system into constructing discrete, multilayered linguistic representations – reminding us that we are neutral between receiving and sending like the Enigma Machine – the way in which this samely-set cognitive system voices out our thoughts in words and sentences, *syntactic structures, morphemes, phonemes*.)

Since the Turing test is to be conducted entirely in a natural language and since our cognitive activities in general seem to go on, mostly, in natural language, Turing's list, in 'Intelligent Machinery', of five avenues for artificial intelligence work included, aside from games and maths, 'cryptography', 'translation of language', and 'language learning'.

To understand why Turing suggests 'cryptography' as a first choice, recall the Enigma Machine's tell-tale redundancies. Some redundancies in intercepted messages resulted from the modular architecture of the machine and key setting system. From masses of these patterns one could determine what this architecture and system were, as a kind of black-box exercise. But this exercise was only possible because of the much more complex, dense, and irremovable pattern of redundancies in natural language, in the plain text German that was enciphered and deciphered by the Enigma Machine. If we analyse the formal pattern of redundancies in a natural language, we may hope to discover much, at a purely formal level, about the linguistic architecture of the human mind/brain that enciphers and deciphers such richly textured transmissions. If Turing could determine the functional innards of the Enigma Machine through analysis of the patterns in its transmissions (speeded by an electronic computing system), why should we not expect a much richer enlightenment in the determination of a most intimate and basic portion of our functional innards, our basic system of encipherment, our day and message settings? Here we have, in essence, the challenge that scientific linguistics has taken up with considerable success, particularly work started up by Noam Chomsky's *Syntactic Structures* (1957) (consult

Chomsky's *Language and Problems of Knowledge: the Managua Lectures* (1988) for an easily accessible presentation of recent work[2]).

Traditional language study mostly had a literary, cultural, and historical bent: anecdotes, eccentric usages, historical changes in pronunciation or particular word meanings, all of it supported by an immense structure of unnoticed common understandings. Scientific linguistics aims to give systematic phonological and syntactical descriptions of the formal properties of the sentences that comprise particular languages *as they are for competent speakers* of them. All this is to be achieved within the explanatory context of a characterization of the child's language acquisition device that accounts for the very surprising number of phonological and syntactical features found in all natural languages and (responsive to 'key' parameter settings) the way these features are structured into a small number of distinctive families, rather than a profusion of arbitrarily adopted conventions. Note the '*as they are for* competent speakers': we do not want squiggles and squoggles. Note also that this attempt to describe a particular natural language without the explanatory context of the others and the child's language acquisition device would be like Bletchley engaged in the hopeless task of deciphering one day's transmissions *without data from other days and complex inferences about the general system of encipherment*.

Let us start with the description of utterances as literal vibrations in the air. Each 'word' is a continuous, complicated stream of sonic vibrations that, in the audiogrammes that map, millisecond by millisecond, the changes in vibration that can affect the eardrum, looks like a smudge-a-squoggle and nothing at all like a discrete sequence of some selection from 25 odd 'sonic-atoms', the simple structural equivalent of the discrete letters that march across this page. This kind of sound description aims at what the linguist would call *observational adequacy*.

In Bletchley's case, what they got was narrow band short-wave activity simply and most adequately described as nothing more than rapid sequences (punctuated by millisilences) of short and long bursts and silence: real 'sonic atoms'. They could say that the messages were in a language whose observable symbols were dot and dash (and blank). How could we determine, on purely formal syntactic grounds, that the message is in a 26-symbol language, not a binary one? The purely formal answer is that there is a vast and compelling horde of patterned regularities that can be most economically described, with most generalizations respected, at the 26 alphabetical symbol level of description. These regularities clearly break into two rule systems, abstract descriptions of the Enigma Machine and the competent German speaker/hearer, respectively (while the only regularities that merit dot/dash/blank description are those few, humble ones that constitute Morse Code).

The first, a wholly 26-symbol affair, is the comparatively simple system that rewrites these individual letters one by one in accord with the complicated addition cipher rule that the Enigma Machine embodies: examining tens of thousands of transmissions one sorts out the contributions of the basic machine, the rarely changed rotor wirings, the daily ring resets, the day and the message initial key settings, as they all contribute to determine how each letter is rewritten. Most distinctively, the 26 letters of this first system *have no individual features, or relationships with each other, other than their (circular) ordering itself*: they are *cyclic numbers*. Functionally speaking, the Enigma Machine has a finite number of internal states whose constitutive modules may all be described by flow chart rules of the form, 'if input is ordinal position nth and you are in the 14th position, add 4, output result,' with most modules confined to simple addition rules for the 26-position cycle, while the progressive rotors introduce a much larger periodicy. Thinking of a language as the output of a rule system or generative grammar, one can regard a letter sequence as a grammatical, or well-formed, Enigma transmission *if and only if* it would have been generated by an Enigma Machine with the assumed monthly and now inferred day setting, and its six-letter initial key indicator. To say it is ill-formed, or 'not Enigmanese', would be to say that the machine must have slipped a cog, or a clerk misrecorded a letter or two, or an electrical disturbance distorted the short wave bursts, or two separate messages were taken as one, etc.

Most significantly and startlingly, the *second* rule system (along with all other natural languages) has *no* rules or relationships of the Enigma form whatsoever. The most direct way to see this is to note that alphabetical order has no linguistic significance whatsoever – if we got everyone to start BCDE and end with A, or switch positions here and there, *the one thing that would be completely unchanged would be the German (or English) language*. But there is much more to the difference than that: natural languages do not use anything like this kind of rule, this kind of vocabulary, nor can they, even in principle, be generated by any sort of finite state machine.

The second rule system presents a vastly different and incomparably richer set of redundancies, though it also, essentially coincidentally, makes use of something like 26 symbols. It is only 'something like' because the graphic 26-letter alphabet inadequately transcribes and digitalizes the complexity of both the phonological and syntactical structures involved; the difficulties we have with spelling immediately suggests this – as does the fool one makes of oneself if one tries to read out loud a passage written in the roman alphabet from an unfamiliar language. Moreover, unlike the Enigma alphabet, natural language letter/phonemes each have distinct individual properties which structure and determine their possible combinations. Further, the formal properties of a natural language *cannot* be described by rules that only make use of these 26 symbols or of sequences of them ('words').

More abstract symbols and more powerful rules are indispensable and essential for generating natural language sentences (and hence for recognizing or understanding or learning them); indeed, essentially *all* the rules of syntax operate on such abstractions. All natural languages present us with a dramatic split between *phonological* symbols, structures, and rules and *syntactical* symbols, structures, and rules. Though the split is not so fundamental in many ways as that between Enigmanese and German, it certainly poses an enormous problem for child language learning, for the child must somehow come up with a vocabulary it never hears *and* acquire complex construction rules in this observationally unavailable vocabulary. (At age three my daughter confidently began a perfectly correct use of the eighteenth-century contraction *amn't* and a year later she insouciantly added the tag formation, 'I am going, amn't I?'; to do this correctly she needed the concepts *verb*, *singular verb*, *auxiliary verb*, *negation*, *contraction*, and *tag negation*, for she has never mistakenly negation contracted a non-auxiliary verb, etc. Even today, at five, she does not know the English words for these concepts she so surely employs, along with the rest of her contemporaries, who exuberantly spout successively more elegant variations on settled adult competence, all at play growing syntax, syntax that they will only much later, and only in the most fragmentary way, hold up for conscious, academically induced inspection.)

Wittgenstein observed that Augustine's 'commonsense' account of language learning makes it seem (paradoxically) as if child Augustine already knew human language and was just picking up the peculiarities of the local dialect. Wittgenstein most clearly stressed the point that words have meaning as their function within the network of procedures that structure our lives, and *not* as mythic lines between thing and word, so that child Augustine's teachers' *point* and utter *sound* procedure is hopelessly inadequate as a general explanation of child language acquisition (the child cannot determine from the pointing what aspect of its appearance or nature we point to, and still less, what function the thing, or the utterance/word, has in our local way of life). But, as we noted respecting the utterance 'gavagai', child Augustine also faces a paradoxically doubled version of this problem in the case of the *utterances* used by his elders.

When the elder purports to draw a line between a thing and an utterance, the line goes both ways: what *utterance* does he point to? What aspects of *the sound structure* determine its phonological and syntactic properties so that the child may recognize when it, or part of it, is said again (by speakers with all sorts of differences in tone, accent, speed, pitch, by the same speaker under a variety of conditions)? If each human language had a completely conventional and idiosyncratic way of individuating speech sounds, the child would seem to face an utterly hopeless task: how to sort arbitrarily sliced sound? Further, since syntactical properties often play a role in determining phonological properties, the child must use notions like *verb, noun, auxiliary, singular/plural, infinitive, tag, subject, direct object*, etc., etc., that cannot be identified with any pattern or aspect whatsoever in the sounds the child hears; further still, the child's native syntactical endowment also requires an intuitive understanding of such concepts as physical object, human intention, volition, causation, goal, and so on.

At the level of audiogrammatically observed sound, utterances are smudge-squoggles which do not *in even a rough way* break down into 25 odd atoms. If you slice the supposed *a* or *v* or *g* sound atom out, you find that it does not look (or sound) much the same as those letter positions in other sequences: moreover, a characteristic portion of the variation is accountable for on syntactic grounds, so you need the grammar to get the 'sound'. We are definitely NOT in the position of the Enigma short-wave receiver clerk, who listens to digitalize into dot, dash, blank so that one or another of the Morse Code alphabetic letters is recognized, though we are more like that clerk when receiving a plain text broadcast in his native language, so that he may automatically use at least some of the structural features of the language to shape, aid, and correct his dot/dash/blank identifications (experienced telegraphers may cease to hear or feel the dots and dashes consciously, and even the letters may disappear into words).

None the less, all natural languages present an enormous number of generalizations at the 25-odd phoneme level. But they are generalizations about how utterances are perceived by competent speakers, *not* generalizations available in the analysis of the physical sound stream, nor in that stream heard by humans who have not acquired the language. In each natural language, oddly enough, each phoneme seems to be a bundle of some 20 *distinctive features*, which can be described as binary choices in physical sound production (tongue touches teeth – yes/no?, plosive puff of air – yes/no?, and so on). Interestingly, hours-old neonates react to sound changes that are phonemic junctures *in some human language or another* while they do not have comparable sensitivity to voice changes that cannot serve this purpose. Somehow, it would seem, the child machine is wired up to listen for sound changes that could be relevant for language acquisition. To further suggest a biological program, the native speaker of Chinese, who cannot hear (or enunciate) the distinction English speakers make between 'flied lice' and 'fried rice', none the less as a neonate of course heard that distinction perfectly well, for it is the maturational experience of acquiring Chinese that eventually wiped out that sensitivity.

Similarly, though a native speaker of English will 'hear' an 'l' phoneme in distinguishing whether someone has said 'split' rather than 'spit', audiographic analysis shows that there is no 'l' sound vibration at all. The English speaker has just heard *the right period of silence* to make the phoneme: the 'l' exists at the psychological level, at the linguist's level of descriptive adequacy, but not necessarily at the level of physical sound. In many other constructions the 'l' will appear as actual sound: the degree and way a phoneme appears as actual sound is determined by a complex interaction of syntactical and phonological rules.

So you see the insuperable difficulties facing the anthropologist who might hope to pick out recurrences of 'gavagai' as a phonological/syntactic structure in the native's language *by listening for recurrent physical sounds*. The task can be achieved, of course, but only through a more or less systematic investigation of the phonological and syntactic system of the native's language which will presuppose and employ general knowledge about natural language and acquisition, about 'the whole system of encipherment'. Since anthropologists are human, they often make use of such tacit knowledge without explicit notice. But the upshot is that in successfully identifying recurrences of 'gavagai', the anthropologist has had to arrive at substantial knowledge of the native's psychology, the native's computational mind set as it embodies his language. *And that computational mind set, when it determines the formal linguistic properties of the sentences of natural language, also determines a core of logical properties, presenting a robust form of linguistic meaning and analytic truth.*

An example will replicate these themes.

1a Tom shaved himself.
1b Dick expected Tom to shave himself.
2a Tom shaved him.
2b Dick expected Tom to shave him.

Notice that, in *1a* and *1b*, *himself* has to be *Tom*, while in *2a* and *2b*, *him* cannot be *Tom* and binds *Dick* or some *Harry* or another supplied by a previous sentence. Speaking linguistically and of all natural languages, a reflexive anaphor such as *himself* is bound in the immediate domain of the subject, while pronouns are free. Notice, however, how these generalizations appear to be contradicted by *1c* and *2c*.

1c I knew the man who Dick expected to shave himself.
2c I knew the man who Dick expected to shave him.

In *1c*, *himself* equals *the man*, not *Dick*, and in 2c, *him* will be *Dick*, not *the man*. This divergence is explained, however, if the human representation of the sentence contains an empty syntactic category, rather like the phonological 'l' of silence in 'split'. So, for the fluent speaker, both *c* sentences have the structure

1/2c I knew the man who Dick expected [*t* to shave himself/him.]

Thus, exemplifying our two generalizations rather than contradicting them, *himself* binds the 'trace' *t* of *the man*, *who*, while *him* is free of *t* (and binds, beyond it, to *Dick*).

Note that the rules that compute this representation are *unconscious*, *untaught*, and *robust*, and in consequence can well be said to be *natural* and *nonconventional*, just the sort of thing that the language acquisition child program *grows in us*, irrevocably setting some parameters, so that our cognitive apparatus perceives the *t* in *1/2c* and the *l* in *split* as robustly as it perceives any other syntactical or phonological feature. Note, also, that the many identifications follow from linguistic form alone. They are *analytic truths*, and bid fair to show us something of the natural structure of our thought, the part carved out in language acquisition through the relentless operations of our language acquisition device, and hence well set apart from conventions and circumstances noticed and learned, from whatever evokes degrees of assent or dissent, flexed by what the whole complex of our experience of the world may suggest.

Note, finally, that the notion of linguistic form here in play is not the metaphysical interface between World, Thought, and Language that *Tractatus* held to be necessary *a priori*; no, it is the upshot of a vast body of linguistic data adumbrated within an explanatory theory of human linguistic perception and processing, and language acquisition. Nor is it the fragile, arbitrary, conventional, *lexical* analyticity found in sentences such as 'oculists are eyedoctors,' 'all bachelors are unmarried males,' 'water is H_2O,' or 'heat is the average motion of molecular particles.'

Professor Quine correctly emphasizes the way in which such conventional and fragile equations can be bent by changes in the 'web of our

beliefs'. If seen in an advertisement in the US, 'oculist' *legally* now
conveys *not an ophthalmologist or optometrist* but someone, without a
doctoral degree, who has some expertise in fitting eyeglasses. A small-
college president who speaks to graduating 'bachelors, masters, and doctors,' might be stating a factual oddity when remarking that 'all bachelors
are unmarried males,' perhaps adding in spare tribute to the perversity
of small samples, 'not like last year, when all three of them were married
females'. Water is indeed H_2O and this was an important discovery
identification, one where we found out what water really was. None the
less, 'it is ice, not water; therefore it is not H_2O' is NOT good reasoning;
nor is 'Galileo believed it was water; therefore he believed it was H_2O.'
It is hard to imagine a more important theoretical identification in the
history of science than that between heat and the mean motion of molecular particles. None the less, if I want heat in my cold hotel room, the
bell captain may expect my hostility if he replies that I already have some
molecular motion. And when we see Hollywood's hard-driving Boss vow
to '*turn the heat up* on that lazy no-good, Jones', and then see him march
into Jones' office, only to exit after setting the thermostat a couple of
degrees higher, we are likely to laugh, rather than to find his action just
what his grim words might lead us to expect. Indeed, we might well say
that *lexical synonymy* is a fragile, accidental redundancy – two words
that differ *only* in phonological properties – one just waiting to be
exploited by wit, fashion, law, science, context, or whatever; by the same
coin, when we discover that two words, e.g. 'Hesper' and 'Phosphor',
name one and the same planet, Venus, we are liable to drop all but one
word from our vocabulary.

So while nonsyntactical lexical meaning and the fragile synonymy it
may afford are easily buffeted by fact or fashion or fortified by scientific
necessity, this is not in the least the case with those natural and robust
analytic truths that result from syntactical, or purely linguistic, form.
This core linguistic knowledge is naturally acquired early and as a system,
not unlike the growth of our visual/spatial faculty, and it is no more
subject to change piecemeal through data input (fact or fashion) than is
the 'hard-wired' central processing unit of a digital electronic computer.

You can fail to see why Turing, or the linguistic scientist, might think
so much of this purely formal approach if you think of the familiar
desk and laptop microcomputers, on which you load, after bootup, one
programming language or another (BASIC, PASCAL, LISP, FOR-
TRAN, etc.) or a particular program written in one of these languages
or in something far more adapted to our computer's specific innards
(something in or closer to its machine language). Couple this with a

confused slide from *the perfectly true claim that*: (1) we, and electronic digital computers, are Turing Machines and, indeed, also Universal Turing Machines, in that all computations can be expressed in the 1, 0, blank Turing Machine tape format *to the conclusion that*: (2) leaving surface sound or visual appearance aside, we might as well describe our thinking in binary notation as any other. Both these plasticities may lead us to think that the structural features of a particular human language tell us no more about the architecture of human cognition than the features of BASIC, say, tell us about the specific architecture of those humans who read and write it fluently, or of the vast variety of digital electronic computers that do so as well, or of both. Fortunately, this grievous misvision does not survive even the cursory examination we gave it.

In fact, even users of micros notice a near exponential difference between machine language and the higher level languages you can nest on top of it. Machine language is, so to speak, the hard-wired, unchangeable, nature of the machine: it is the most specific level of abstraction at which we can exploit the information-bearing power of the circuitry. When we load one or another more abstract programming language, we create a *virtual machine* by translating its properties into the machine language below, naturally at some cost in processing capacity: things go faster and take up less space in machine language. Since elegantly simple and natural translations of the *PM/TM* style would in practice prove prodigally demanding, a term arose for a clever, ad hoc maneuver that would exploit some shortcut machine language stand-in that would do the higher level job without the expense of full *PM/TM* style translation: *hack* (as in *clever hack* or *hacker* (MIT, 1960s)). I think of the red button on the back of my first electronic calculator as a neanderthal hardware hack, the equivalent of shock therapy for humans caught in pathological cognitive loops. Turing sketched a more sophisticated hack in his provisional solution for the bridge collapse problem he had discussed with Wittgenstein.

The processes of inference used by the machine need not be such as would satisfy the most exacting logicians. There might for instance be no hierarchy of types. But this need not mean that type fallacies will occur, any more than we are bound to fall over unfenced cliffs. Suitable imperatives (expressed *within* the system, not forming part of the rules *of* the system) such as 'Do not use a class unless it is a subclass of one which has been mentioned by the teacher' can have a similar effect to 'Do not go too near the edge.'

But this solution, reminiscent of St Paul, would seem to provide a telltale clue for Turing testers. (Unless, remembering our own case, we imagine that the child machine will develop two cognitive language systems, a more basic one in which inhibitory commands will be received and followed out, in which threatening ambiguities and strange loops will be recognized and evaded *and not allowed to come to disturb the attention of* a second cognitive language system in which the Turing test, and conscious conversation and thought generally, will be carried on.)

To solve a portion of Plato's problem about mathematics, Chomsky speculatively ascribes a much grander and more comprehensive hack to evolution.

How did the faculty of number develop? It is impossible to believe that it was specifically selected. Cultures still exist today that have not made use of this faculty; there language does not contain a method for constructing indefinitely many number words, and the people of these tribes are not aware of the possibility of counting. But they certainly have the capacity. Adults can quickly learn to count and to do arithmetic if placed in an appropriate environment.... In fact, the capacity was latent and unused throughout almost all human history. Plainly it is not the case that people who could count, or who could solve problems of arithmetic, were able to survive, so the capacity developed through natural selection. Rather, it developed as a by-product of something else, and was available for use when circumstances called it forth.... It is possible that the number faculty developed as a by-product of the language faculty, [which] has features perhaps unique in the biological world.... Human language has the property of discrete infinity, and the same is true of the human number faculty. In fact, we might think of the human number faculty as essentially an 'abstraction' from human language, preserving the mechanism of discrete infinity and eliminating the other special features of language.

If Chomsky is right, this might fill in part of the explanation for the gaps Wittgenstein notes in our everyday accounts of rule following, of the *natural* as opposed to *mathematical* as Wittgenstein uses the contrast in discussion with Turing. We are, you might say, our rules.

If one is still tempted to think that linguistic form is somehow mostly just the product of the architecture of the human articulatory system and hearing (like the physical contingencies of short-wave broadcast and reception), human sign languages form a most powerful counterexample. The born deaf child whose first language is American Sign Language computes syntactical representations with very much the same features as found in natural languages. Even counterparts of distinctive features and phonemes are found, as if the formal, purely cognitive properties of language drive and mold the perceptual articulatory aspects of language (ASL, whose known origins go back to eighteenth-century France, has an inflectional system more the style of Latin than English).

Turing remarked that neither eyes nor ears might be essential in that Helen Keller acquired fluency in English though made completely deaf, blind, and dumb by an infection in her nineteenth month; when six years old, Anne Sullivan began to teach her through touch, and Keller went on to graduate with distinction from Radcliffe College and to write a number of books. In the 1970s, 'Genie' tragically exhibited the opposite case: her father isolated her from all exposure to human language from age 16 months to nearly 14 years of age (well into puberty and well past the critical period for language acquisition). From this point on, Genie fairly easily picked up rough use of several score words but was, even with years of the best teaching, wholly unable to grasp the simplest syntactic structures, wholly unable to produce or understand phrases or sentences, though her visual intelligence, for example, as exhibited in pattern and analogy recognition, proved distinctly above the human average. Genie's case is not unlike that of children, born with vision-blanking cataracts, who only have them removed too late, after their first, and critical, year: such children are sadly unable to make use of sight, they cannot effectively see the array of objects in the retinal irradiation just as Genie cannot ever hear the sentences, the thought-forms, in the noise.[3]

Imagine a change in our frontispiece. Replace the mechanical hand with a chimpanzee one. Ironically, in that human signers and the study of human sign language have suffered greatly from prejudice and inattention, one stimulus to the large-scale research of the past two decades on human sign language was the late 1960s attempt by R. A. and B. T. Gardner to teach Washoe, and other chimpanzees, several score of ASL signs. While daily training in their early years will teach such a gestural vocabulary in chimpanzees, they saliently fail to show any of the syntactical or productive features that make ASL, as embodied in fluent human signers, like a natural language. Chimpanzees do not make sentences and even the individual signs they do approximate lack the clear cut 'enunciation' and system of distinctive contrasts that characterize human signing. Extravagant claims about chimpanzee signing abated when Herbert Terrace, who set out to replicate the Gardners' work with a more concerted effort to establish something like sentence formation, reversed himself after four years of work with a chimp he had originally defiantly named Nim Chimpsky; Nim, Terrace finally concluded, in fact had not been able to pick up any sentence-forming capacities. The small amount of apparently contradictory evidence was often, he argued, and in some instances demonstrated photographically, the result of inadvertent cuing by human investigators.[4]

Geneticists estimate we share 99.5 per cent of our genes with the

chimpanzee, who is also in physical structure and social behavior hauntingly familial (indeed, researchers have achieved chimp/human conception and normal embryonic development, though these experiments have
been halted, deliberately, long before birth). I felt immediate rapport in
my first hours with one of the Gardners' chimpanzee students, a seven-
year-old female, Moja. Maintaining frequent, confident eye contact, she
deftly spoon-fed me part of her breakfast (but not her favorite part,
bananas), cleaned my teeth of plaque without damaging my gums, and
manicured my fingernails, later successfully inviting me to nap with her
in the 'spoon' position, and still later using me to establish a relationship
with a two-year-old male chimpanzee, Loulis. When Moja pulled a woolen
cap down into a ski mask, I remembered that the chimpanzee is the only
animal other than us who will recognize, in a mirror, that someone has
painted an odorless white spot on her forehead while she was anesthetized;
from the time she wore the mask, I carefully kept away from her teeth.
Yet I am soberly conscious how rightly and sadly that primatologist spoke
who said to me, 'I know now that I am never going to have conversation
with a chimpanzee.' Let us give Turing a penultimate word here.

It clearly would not require any very complex system of genes to produce an
unorganized machine. This should be much easier than the production of such
things as the respiratory centre. This might suggest that intelligent races could
be produced comparatively easily. I think this is wrong because the possession
of a human cortex (say) would be virtually useless if no attempt was made to
organize it. Thus if a wolf by a mutation acquired a human size cortex there is
little reason to believe that he would have any selective advantage. If however
the mutation occurred in a milieu where speech developed, some selective advantage might be felt.[5]

Turing spent much of his last three years on mathematical theories of
neurological and biological development ('the idea is that by different
trainings certain of the paths could be made effective and the others
ineffective . . . so as to grow to form a particular circuit . . . The
brain structure has to be one which can be achieved by the genetical
embryological mechanism, and I hope that this theory that I am now
working on may make clearer what this restriction implies.'). Indeed, his
1952 Royal Society paper, 'The Chemical Basis of Morphogenesis', was
acknowledged by Ilya Prigogine as an anticipation of his own Nobel
Prize-winning work in 1972,

The development of irreversible thermodynamics of open systems by the Brussels
school had, by the 1950s, led to the investigation of non-linear processes. . . . It
was only then that we noticed a remarkable paper by A. M. Turing (1952) who

had actually constructed a chemical model showing instabilities. His work had previously escaped our attention because it dealt with the more specific subject of formation of morphogenetic patterns. The work we have undertaken since then has demonstrated the relationship of this type of behavior to thermodynamics as well as its wide applicability to biology.

Turing's biographer rather drily adds that Prigogine had apparently forgotten that, as a visitor to the Manchester University Chemistry Department where Turing had just delivered a seminar on his morpho- genetic theory, he had spent much of 29 February 1952 in a reportedly lively interchange with Turing about Turing's theory.[6]

In identifying natural language sentences with their structural form as perceived and processed by competent human users, we replay the theme of Turing's 1937 paper and Wittgenstein's *Investigations*. We do so as well in stressing the gross inadequacy and incoherence of the common- sense account of child language acquisition that likens it to adult learning of a foreign language; we need, ultimately, a biological account of langu- age acquisition and function, which is to say a physical and mechanical one as well. As Chomsky says,

When we speak of the mind, we are speaking at some level of abstraction of yet- unknown physical mechanisms of the brain, much as those who spoke of the valence of oxygen or the benzene ring were speaking at some level of abstraction about physical mechanisms, then unknown. Just as the discoveries of the chemist set the stage for further inquiry into underlying mechanisms, so today the discoveries of the linguist-psychologist set the stage for further inquiry into brain mechanisms, inquiry that must proceed blindly, without knowing what it is looking for, in the absence of such understanding.

Turing listed translation as a less formal, and therefore less immediately promising approach to language. Translation from one natural language to another raises the nonformal issue of meaning (semantics). In the late 1950s and early 1960s, artificial intelligence enthusiasts were often as optimistic about the speedy achievement of machine translation as they were about chess mastery and 'problem-solving': a few years would do the job. In fact, the attempt to approximate the achievement of a reason- ably good human translator through computer machinery proved a most notable failure (though not with anything like the relative eventual success of chess-playing machines, several decades have seen some considerable progress). At the simplest level, machine translation amounted to little more than something like a computerized English/Russian dictionary: you feed a Russian string of words in, and each is successively replaced by an English word (that has 'the same meaning'). It is worthwhile to

consider why this most attractive and simple proposal fell resoundingly
flat (though, again after several decades of work, there has been some
success).

In the first place, while there are syntactical universals and constraints,
some differences in parameter settings can determine a very considerable
difference in the syntactic structures between languages. Recent linguistics assumes a sentence has an underlying structure that can be described
through labeled brackets. E.g., 'Jill hit Jack' would be

$$[\ \ [\ \ [Jill] \ \ ] \ \ \ \ [\ [hit] \ \ [Jack] \ \ ] \ \ ]$$
$$n \ np \ \ \ \ \ \ \ \ \ v \ \ \ \ \ n \ vp \ \ \ \ sentence$$

English takes Jill as subject and Jack as object; English is a SVO language.
But there are VSO and OVS as well. In addition to the phrase structure
rules there are also transformational rules that rearrange or delete material
(e.g., the rule that left the trace *t* by moving *the man, who* in 'I
knew the man who Dick expected to shave himself.'). The deletion
and compression produced by transformational rules causes a syntactical
ambiguity in 'Flying planes can be dangerous' in that the words convey
either that piloting is a perilous occupation or that living near an airport
puts one at risk; you can see that the difference is *syntactical* if you
disambiguate from the infinitive *can be* to the finite *is* or *are*. While
much of phrase structure seems common to all natural languages, transformational structure may be more local.

These facts alone rule out the simple computerized dictionary
approach. On the other hand, they hold out hope for a more complicated
approach which would take full syntactical structure into account.
Roughly, the translator might be programmed to 'de-transformationalize'
sentences from Russian into underlying phrase structures with appropriate labels (*verb, verb phrase, subject*, etc., are, after all, universal); then
it would 're-transformationalize' using English rules, and finally, from a
more carefully designed Russian/English dictionary, it would insert the
nearest English equivalent for the words or word parts in the Russian
sentence.

Proverbially, an early machine translation from English to Russian and
then back again is said to have transformed 'The spirit is willing but the
flesh is weak' into 'The whiskey is OK but the steak is rotten.' The more
sophisticated syntactical engine + dictionary I sketched would at least
avoid this kind of foolishness in that the *is willing* requires that the
subject be a person, so the program would handily eliminate the 'alcoholic
beverage' reading under the dictionary entry for *spirit*. The sentence
reads less well if 'the flesh is weak' begins it (perhaps also changing *but*

to *though*) because *is weak* does not so clearly, and syntactically, require a personal subject; moreover, 'weak' rides the continuity between body and mind, between physical and personal, at its biological interface – weak memory seems little more spiritual than weak kidneys, and *weak resistence* can be as flatly true of an immunological as of a military fortification system (a minor academic industry has arisen to mock the bare idea that computing machines could ever think; no one denies them viruses). Even 'spirit' arose from the idea of respiration, of prana, of vital principle or inner spark, before it took on the sense of soul or supernatural, and one has low spirits, good spirits, even broken spirits: medieval pharmacy contributed both the notion of essence or active principle of a substance, particularly when extracted in a liquid form (hence, spirits of ammonia, spirits of turpentine, or spirits of wine) *and* of any such extraction if achieved by or stored as a liquid in alcohol. Obviously, a machine translator that tries to take account of all of this will require great complexity; the dictionary will begin to become an encyclopedia.

Worse, since languages pack quite various assortments of senses into particular words, which in turn resonate with each other through a range of syntactical levels and phonological levels, we are quite unlikely to find any translation that will carry over anything like all these properties. As a practical matter, human translators have long been only too aware of this. The novelist, C. S. Forester, has a talented forger, Dr Claudius, analyze the problem of composing a fake letter designed to spark a Napoleonic War sea battle.

Have your letter *composed* by a Frenchman. You gentlemen may pride yourselves on writing good French, grammatical French, but a Frenchman reading it would know it was not written by a Frenchman. I'll go further than that, gentlemen. Give a Frenchman a passage in English and tell him to render it into French and a Frenchman will *still* be aware that all is not well when he reads it. You must have your French composed ab initio by a Frenchman, contenting yourselves with merely outlining what is to be said.[7]

Idiom is one of our terms for some of the most flagrantly difficult, and unpredictable and unmanageable, examples of this. Idioms might well be called *clever hacks* in that in them the phonological structure, superficial syntax, and underlying structure *all* resonate together in happy idiosyncrasy, each unexpectedly contributing to terse and taut expression. Our more sophisticated program will suffer massive breakdown here, for after you have stripped off the phonology and 'de-transformationalized' the syntax, you will have already lost most of the meaning of the phrase and indeed will inevitably mistake what is left. Idiom lists can help us to avoid wildly mistranslating particularly well-known cases, but these are just extreme and extremely familiar instances of a general phenomenon. As Forester's forger wisely suggests, we often must content ourselves with *outlining* what is to be said. We cannot expect a formalized procedure to produce uniformly good results: rather, we may have to seek, pragmatically, for what will achieve the desired effect, taking into account whatever linguistic and nonlinguistic factors seem relevant to our particular circumstances.

Two more theoretical points are suggested by this practical one.

One is the failure of the central claim of the 'Generative Semantics' movement that briefly flourished in the 1970s. Several linguists and psycholinguists championed the hope that all natural languages had the same ultimate underlying syntactical/semantical structures (the English word 'kill', for example, was supposed to decompose into the universal semantic structure CAUSE + TO BE + NOT + ALIVE, with the capitalized letters representing something common to all human languages). Another is that it makes reasonable sense to say that we think in our natural language, that it is our machine language, the one we shall so often translate into for naturalness and ease of handling (even though this means a loss in what logic teachers call 'elegance' and 'simplicity').

But although we undoubtedly do much of our more discursive and sequential thought in natural language, one of the most striking and well-supported results of recent work is support for the *modularity of mind*. The empiricist picture of perception fading into thought and thence to memory from whence it may by association re-emerge as thought is more subject to disconfirmation when it reappears as the supposition that there is something like a central processing unit (attention and executive consciousness, so to speak) and a general memory store, through which what we cognize is manipulated; and with this, we also find the view that Learning and Cognitive Development are equally general, incremental processes. To the contrary, we appear to have several native faculties and subfaculties, each with a critical biological development period, with particular brain locations and neurological features, with particular kinds of representations and memory forms that are wholly or partly unavailable to consciousness; we eke out, or piggyback, on these modules to hack away at the many problems we confront for which we are not specifically well designed.

It is no great surprise that we have a visual/spatial perception module and we may easily extend this sanction to separable visual thought and memory. We all know that *a picture is worth a thousand words* under- and overstates. It overstates in that for us words may often be indispensable or at least more powerful and economically expressive than pictures. It understates in that visual representations are often necessary or at least enormously more efficient for our cognition. While in principle, as Turing showed, we can convert any input picture into a binary sequence, and vice versa, we humans natively settle for a few, mutually exclusive, specific and quite discrete representational modalities.

It is genuinely startling to learn that our retinas stimulate two visual systems, one running through the *superior colliculus*, the dense and richer other running through the lateral geniculate nuclei; damage to the second, which seems to support the more sophisticated and geometric features of our vision, can cause 'blind sight', a condition in which the subject cannot consciously see but none the less can successfully reach out for particular objects at the experimenter's request. It is equally startling to learn *facial recognition* is quite separable from general visual perception, with its own maturational program, representational structures, and neurological location; while we all know how extraordinarily difficult it is to describe another's face in words, it is a shock to learn that specific brain lesions may suddenly render a human incapable of visually recognizing family members or friends, though he can see perfectly well and indeed can do sketches of them that others can recognize. We might well add to Turing's metaphor that we are *cloved* onions. As we strip the outer layers, we find evidence of separate cognitive organs whose unity is often, and disturbingly so, skin deep.

11

Stories of Consciousness

Man is so complicated a machine that it is impossible to get a clear idea of the machine beforehand, and hence impossible to define it. For this reason, all investigations have been in vain, which the greatest philosophers have made a priori. Thus it is only a posteriori or by trying to disentangle the soul from the organs of the body, so to speak, that one can reach the highest probability concerning man's own nature.

La Mettrie

Turing's observation that machines will go into endless loops when trying to predict their own behavior suggests that a sufficiently complex machine might also come to suffer from that seemingly inevitable human delusion: believing that one has free will and is able to make choices that transcend physical law.

Hofstadter

As maturation continues, the behaviors that these separate systems emit are monitored by the one system we come to use more and more, namely, the verbal, natural language system. . . . The mind is not a psychological entity but a sociological entity, being composed of many submental systems. What can be done surgically and through hemisphere anesthetization are only exaggerated instances of a more general phenomenon. The uniqueness of man is his ability to verbalize and, in so doing, create a personal sense of conscious reality out of the multiple mental systems present.

Gazzaniga and LeDoux[1]

Extending to humans the findings of animal experimenter Roger Sperry, Dr Joseph Boden, beginning in 1960, markedly reduced severe epileptic seizures in several patients by severing the *corpus collosum*, the nerves that connect our two cerebral hemispheres. By his work, and from subsequent research by Michael Gazzaniga and others, we appear to have learned, and confirmed, something about the nature and place of consciousness in our cognition. Our bilaterally symmetrical bodies exhibit a crossover when it comes to the brain. In most humans, the right side of the brain sends motor output to, and receives sensory input from, the *left* side of the body; while the left hemisphere controls and 'senses' the right side. (Our eyes, however, are each split down the middle of the visual field.) Linguistic abilities are normally largely localized in the left side of the brain; certainly, it initiates and controls speech. The first noted major effect of the collosectomy was that *both* sides of the brain carried on independent cognitive activity: so much and startlingly so, indeed, that some commentators said that there were now two persons while others found the left, verbally controlling hemisphere a person and the right, a dumb automaton. For example, one might place an object, say a pipe, in the subject's left hand (outside his visual field). If asked *what is in your left hand?*, the subject (that is, the left side of the brain) comfortably fabulates a response of, say, *pencil* or *compass*. When, however, you ask *use your left hand to pick what you had in it from these pictures*, that unschooled appendage points to the pipe picture, for the right brain communicates with the left hand.

Another notable feature is that the left side of the brain confidently denies any of this independent right-brain activity and smoothly rationalizes this denial. Michael Gazzaniga consistently obtained this result with a number of patients in countless trials. As a typical example, a patient was asked to pick which picture cards fitted with a briefly flashed picture. Her left brain saw a chicken's foot and her right hand reached for the card with a rooster's head; her right brain, however, saw a snow-covered car and driveway, and her left hand picked a card with a snow shovel. When she was asked why she had selected a shovel, the patient smoothly responded that it was for cleaning up the chicken droppings. A patient whose right brain saw the instruction 'rub' rubbed his right hand with his left; when asked what the command was, he smoothly replied 'itch'. When the instruction was 'assume the position of the flashed word' and the word 'boxer', the patient assumed a pugilist posture and when asked the command, he replied just as smoothly but more accurately, 'boxer'; if, however, his arms were restrained, he easily (but mistakenly from his right brain's viewpoint) replied that no command had been flashed.

Most interestingly, some patient's left brains could pick up emotional reactions of their right brains. P. J., an adolescent boy, blurted out to male personnel, 'Hey, no way, no, way, you gotta be kidding' when his right brain was flashed the command 'kiss'. Yet he could not specify what it was that he would not do (his initial reaction was identical when the word was projected to his left brain, though he then could specify what he was rejecting). Using a scale running from 'like very much' through to 'dislike very much', P. J.'s left brain would give much the same emotional ratings for words when they were flashed to his right brain, though he could not at all say what the words were. (Interestingly, the one significant exception was 'Nixon' (then President), who got 'like' when the left brain could read the word but 'dislike' when it just had the right brain's emotional reaction to go on; 'Dad' and 'God' dropped from 'like very much' to 'like' and TV moved up from 'like' to 'like very much'; 'school' and 'police' rose from 'dislike' to 'undecided'; for six other items there was complete agreement.)

P. J. proved in one respect exceptional in that his right brain was able to exhibit substantial linguistic abilities and a sense of personal consciousness as opposed to the more limited, reactive behavior found in other split brain cases (which Sir John Eccles labeled as 'automatisms'). P. J.'s right brain could use his left hand to spell out correct answers to the sort of questions we usually ask to determine whether someone is 'all there': who are you?, girl friend?, favorite person?, what day is tomorrow?, and so on; the one deviation was that the right brain spelled out 'automobile race' as the job he would pick, though his speaking self often said that he wanted to become a draftsman.

Significantly, Gazzaniga and others have been able to obtain similar results in surgically untouched, nonepileptics through anesthetizing the right or left hemisphere. Given the dubious history of psychosurgery, I might add that collosectomy seems to cause little or no damage to everyday cognitive, sensorimotor, or emotional function; only mammals have *corpus collosuma* and these hemispheric connections play a particularly rich role in primate visual functioning. Indeed, the revealing results can only be obtained under highly restricted experimental conditions. In normal vision, for example, our eyes are rapidly moving so that all visual information is directly available to both hemispheres; and, of course, if I wonder what my left hand is doing, I can take a look at it (note however, that if one of your hands has felt around fairly similar objects in a bag and picked one out, it will upon return be able to find the same object, or its twin, while your other hand will not be able to do so).

Gazzaniga argues that our verbal module supports, indeed is, our consciousness and personhood. But he stresses that much of what it does is to explain, unify, and narrate the collection of inward and outward activities that our many modules produce. Since we have had quite some time to observe ourselves, to explain ourselves to others and ourself, we naturally do a good, or at least complicated, job of it – this is our common human curriculum, our perilous, madcap act of self-creation, thrown upon us in our first sentences, when we start the stories we shall only learn stage by disconcerting stage to stand back from and tell better, until our last words. We have been passing the Turing test.

The familiar analogy, 'I can read him like a book' (along with its pedestrian communications' miniature, 'Do you read me?') suggests not only a paradoxical paradigm of intelligibility but also an intriguing conception of what it is to understand a human being, what it is to find out what it is like to be a particular human being, what it is like to have consciousness and a subjective existence. If you are particularly transparent – your real motives and your inner life unusually, even naively and vulnerably, open to view – then you meet the standard of accessibility established by written narratives, whose leaves of consciousness may be turned by any eager reader. More deeply, what goes on in reading a narrative is structurally analogous, or even identical, to what goes on in understanding, in grasping, what it is like to be a particular human being, whether a stranger, a friend, or most intriguingly, oneself.

What does seem common and central to consciousness is the ability to carry on a monologue or dialogue about oneself and others in the intentional idioms of a natural language, to be able to give life narratives (and updates) about oneself and others, to be able to distinguish, where possible, how one feels from how one appears or pretends to feel, to be able to give a fair account of oneself and what one knows and what one has been through.

Of course, our pronomial idiom continually intimates to us the message of Descartes' *Cogito ergo sum*: that consciousness is instantaneous, locationless, indivisible, and transparent, like an interior room whose light bulb is either brilliantly on or pitch black. Certainly the particular version of the intentional idiom Descartes used continues to beguile us so in his *Meditations*, whose self-narrative is one of the most powerful, compelling, and convincing ever written. But, as the lengthy, dialogic character of the 'one, sure' Descartes/Turing test immediately shows us, a single flash, a single interior or exterior mouthing of words, shows us nothing. Descartes did remark that it would be easy enough to make a human-looking automaton that responded with 'I am in pain' whenever we touched it, and that, Descartes asserted, does not in the least show us that the mechanism thinks or is conscious. But, if that is true, why should it make any difference if the automaton says (loudly or subvocalizingly) 'I think; therefore I am'? What makes the sentence significant is that the creature mouthing or minding it has the secure ability to go on, responding appropriately to questions we or the creature might ask itself ('secure ability' is a blank cheque for the cognitive competences realized in concrete mechanisms that make the ability real as opposed to miraculous or coincidental).

Someone may say, 'either the interior light is on, or it is not – that is all there really is to it – and we can never know, respecting another, whether it is or is not!' But this is hopelessly blatant and empty-minded homunculizing. It says nothing and it tells us nothing of the thing within who looks about the cozy, well-lit room – and why should not the thing

within be a good navigator in the dark or blind, even an heroic Helen
Keller of the inner room, managing to will and think in echoless darkness?
*What makes the Cogito mean something is that Descartes has already
embedded, and can go on to further embed, this sentence within a self-
questioning dialogue and self-narrative that makes it powerfully clear
that Descartes passes his own test.*
    I hope I have ours.

Why were Democritus and Descartes, and still more La Mettrie, Russell,
and Turing persecuted in such a hauntingly similar way – and indeed
felt to bring it on themselves by goatishness, by (somehow) mocking,
obscene, gleeful conduct. Classically, Democritus was the 'Laughing
Philosopher'; a few lines from his 60 books of experimental science
survive, while apart from Aristotle's summaries we have spiritualist con-
demnations of his work. St Paul's leap from paradox to calumny ('evil
beasts, swollen bellies') is representative of the ancients and the medievals
and persists today. The seventeenth century set the image, in words and
pictures, of Descartes at dissection; his arch eyes hold ours as he displays
the brain and ocular equipment, having removed several of Turing's
onion skins. The only picture I know of La Mettrie gives us the frivolously
grinning, fleshy gourmand. Twentieth-century newspapers have dis-
played Russell's caricature as cocky and mocking, ridiculously apish or
goatish in old age.
    Turing, *et al.*, insist on a painful pun, a disquieting identification. We
are prone to picture them as *naughty*: ourselves unclothed, we haughtily
insist that only the childish, the callow village atheists, could be so rude.
G. K. Chesterton, who is remembered for his Father Brown detective
stories, exhibits the venom that powers this malignant projection when
he mocks a much better, and profoundly more moral, writer, Thomas
Hardy.[2]

Hardy went down to botanize in the swamp; he became a sort of village atheist
brooding and blaspheming over the village idiot.

We *are* machines, and the persons that we are also, they are stories,
accounts, dreams: though they are (largely) true stories, fair accounts,
and waking dreams, negotiable, shared. Our faces are flesh, are masks,
that we may imagine ourselves able to doff, like the many films in which
a character is unmasked as circuitry, as mere robot. Yet were the same
literally to happen to a human being (to, for example, the human actor
who pretended to robothood) we should not find a second, and real, face
underneath the mere epidermal, adipose, and muscular tissue: we find

machine within, our expressive face finally unmasked by our intelligent machinery, only theoretically sluffed for feature and pose are practically indispensable, along with the many dream faces within. Shakespeare was right – or right enough for *us*.

The birth of the intelligent computing machine is equally the full, final unmasking of the human as a thinking device, a biological mechanism. Those who followed in Descartes' wake wondered whether the world of automata was their dream; we have awakened to discover that we, as persons, are the dreams of human machines.

Notes

Preface

1 Thomas Kuhn, *The Structure of Scientific Revolutions* (Chicago: University of Chicago Press, 1970), p. 94.

2 Otto Neurath's remark that we are sailors who must do ship repair without recourse to drydock is quoted in the original German by W. V. O. Quine at the beginning of his *Word and Object* (Cambridge, Massachusetts: MIT Press, 1960), p. x; for Neurath in an English translation, see A. J. Ayer, ed., *Logical Positivism* (New York: The Free Press, 1959), p. 201. I criticize Quine's views about language, translation, and cognition in chs 6 and 10. Quine regards all of language as social and conventional, as learned and consciously available; hence, all of it is up for revision and the beliefs we hold *in* it constitute a web, so that no belief is exempt from influence through changes in others. To the contrary, much of natural language is innate, automatic, unconscious, or becomes so by native acquisition in our first few years, while Quine's comments do have application to less structural, more superficial features of language. Quine's views were supported by the behaviorist linguistics that flourished, in linguists' theories though not their practices, through the 1930s into the 1950s in the United States. He also very much depends on their view that language can be described as an actual physical stream of sound: that a phoneme can be individuated as physical sound. This view is simply no longer supportable in scientific linguistics.

1 Shake Hands

1 Douglas Hofstadter's *Godel, Escher, Bach: an Eternal Golden Braid* (New York: Basic Books, 1979), which introduces the term 'strange loops', teaches one much about paradoxes and related phenomena, whether logical, linguistic, mathematical, pictorial, or musical. There is a vast literature about intention and intentionality, about persons and what makes them persons, and about the

idiom we use when talking about persons when they are being persons; Daniel Dennett's *The Intentional Stance* (Cambridge, Massachusetts: MIT, 1987) provides a critical look at it.

2 Descartes' assumptions about the similarity between dreams and waking existence has proved remarkably prescient. We know today that when we dream we exhibit rapid eye movements (REM sleep); not only do our eyes move about as if we were seeing, the rod and cone cells of our retinas fire off neurologically as they do when light strikes them in everyday 'lids open' vision. We have yet to establish to what degree there are correspondences between the patterns of retinal firings, dream content, and eye movements. In normal waking vision, our brain constructs and fills in an experience of a three-dimensional world of colored objects from some fairly meager two-dimensional retinal input (and we know rapid eye movements are a necessary part of this process), so it may well be that what we *see* in our dreams are visual construals initiated and fueled by coincidences in essentially random retinal firings, which are then given some narrative interpretation and even plenty of the beginnings of co-ordinated motoric responses, though of these, all but those that fire off the eye movements are (fortunately) almost entirely inhibited from getting to our muscles. Similarly, visual perception, visual memory, and visual imagination make use of much the same neurological staging areas and communicative pathways. Having the same neurological system serve as camera and projector, receiver and creator (hearer and speaker) is a typical economy of nature; computer hard- and software engineers have found themselves driven to similar economies.

3 Stephen Gould's *The Panda's Thumb* (New York: W. W. Norton, 1982) makes the opposite point: since opposable fingers are most useful, given we do without the usual mammalian premium on speed in four-legged running about, nature will shove what it can into this function. This point is also applicable to computer development.

2 The Classical Agenda

1 Anya Hulbert and Tomaso Poggio, 'Making Machines (and Artificial Intelligence)', in Stephen R. Graubard, ed., *The Artificial Intelligence Debate* (Cambridge, Massachusetts: MIT, 1988). The debate of Graubard's volume, between connectionist and programmatic approaches, is prefigured in Alan Turing's work.

2 In *The Black Athena* (New Brunswick, New Jersey: Rutgers University Press, 1987), Martin Bernal challenges the priority of the Greeks respecting the alphabet and indeed claims many achievements usually credited to pre-Socratic Greeks as imports from an Egypt that is rather more Nubian and less Semitic than is usually assumed. Nineteenth-century northern European scholarship did falsely romanticize an ancient Greece decidedly pure of all but northern European influences. The initial reaction of professional historians is that Bernal strains hard beyond and against the available evidence in the opposite

direction. What has long been simply indisputable is that early Greek science
and philosophy began in cities on the coast of what is today Turkey.

3 The Gathering Storm

1 Julien Offroy de La Mettrie, translated by G. C. Bussey, *Man a Machine*
(Chicago: Open Court, 1912), pp. 140–3. This volume also contains selections
from *The Natural History of the Soul* and Frederick's Eulogy. The translation
is literal and wooden, and sometimes flatly wrong. We also have Aram Vartanian, *La Mettrie's L'Homme Machine: A Study in the Origins of an Idea*
(Princeton, New Jersey: Princeton University Press, 1960). Since Vartanian
was perfectly aware that the Bussey translation was unsatisfactory (indeed, it
was long out of print when Vartanian's book was issued) and only preceded
by some now extraordinarily rare, 'popular' translations printed in 1749 and
1750, one is unavoidably reminded that, at least prior to the 1960s, 'obscene'
texts could be safely published and distributed in the US *if* they were in a
foreign language, preferably a dead one (similarly, translations of classical
Chinese love novels into English safely rendered the more explicit passages
into Latin). I am sure La Mettrie would laugh to see that his words can still
evoke such protective measures, though there might be a touch of bitterness
in his laughter. In his *History of Materialism* (New York: Humanities Press,
1950), Friedrich Lange comparatively defends La Mettrie's personal life in
Section 4, p. 79, also observing that there had never been any specific accusation
of personal misconduct against La Mettrie. 'But La Mettrie's death in a delirious
state after devouring of a large pâté aux truffles is an object that so completely
fills the fanatic's narrow horizon as to leave room for no other idea' (p. 91).
Frederick the Great comes out well in all of this. He anticipated that La Mettrie
would not be safe in the Netherlands and he offered him protection and an
honoured place some months before La Mettrie could see the extreme danger
that faced him. In Frederick's court, La Mettrie could talk with the King in
a familiar and friendly manner; this perhaps ruffled the feathers of Voltaire,
who was much older and more widely honoured and accomplished, for Voltaire
joined in the revolting personal attacks on La Mettrie after his death. La
Mettrie insisted that his attending doctors treat him with a course of bleeding
and cold baths and Frederick believed that this contributed to La Mettrie's
failure to recover. Anticipating Professor Lange's charge of frivolity, Frederick
obviously wished La Mettrie had not issued his short books on happiness and
sexual pleasure, but, as the Eulogy makes clear, Frederick blazed out an
aggressive defense of La Mettrie's *L'Homme Machine*, his medical work, and
his person.
2 There are now a number of accounts of Lady Ada Lovelace's life and some
dispute the degree to which it is reasonable to call her 'the first computer
programmer'. See Dorothy Stein, *Ada Lovelace and the Thinking Machine*
(Cambridge, Massachusetts: Harvard University Press, 1985).
3 Holmes' clue of the dog who did not bark in the night appears in 'Silver

Blaze'; heavily backed in an impending race, the favorite horse, Silver Blaze, disappears, his trainer found a few hundred yards from the stable, killed by a 'blunt instrument'. Since the fierce guard dog did not bark in the night, Holmes deduced that the dog knew well the person abducting Silver Blaze (the trainer himself was in fact the only person available for that role; having snuck the horse off to nick a sinew, the trainer provoked Silver Blaze to kick his head in). Holmes often remarked that humans are not naturally given to such 'backward reasoning' (*modus tollens* is, of course, the logician's term for it); it has proved a powerful mode for the electronic digital computer.

4 Russell's praise of Peano's symbolism is in his *Autobiography, Vol. 1* (London: Unwin Hyman, 1967), p. 218.

4 Meaning Must Have a Stop

1 Gödel's comment about Turing is from his 'On Undecidable Propositions of Formal Mathematical Systems', in Martin Davis, ed., *The Undecidable* (New York: Raven Press, 1967), p. 72. His incompleteness results first appeared as 'On the Formal Theoretical Advances of the *Principia Mathematica* and Related Systems', *Monatshefte für Mathematik und Physik, Vol. 38* (1931). The Davis volume contains both of these papers as well as Turing's 'On Computable Numbers, with an Application to the Entscheidungsproblem', *Proceedings of the London Mathematical Society*, No. 42, 1937, and the corrections of the next issue; there also is Alonzo Church's 'A Note on the Entscheidungsproblem' and a posthumously published paper in which Emil Post achieved, at the same time, the undecidability result of Turing and Church by yet another route. The convergence of three quite different formalizations of the intuitive notion of computable, each arrived at quite independently, is impressive. Turing had completed but not published his work when Church's note was published. However, there is universal agreement that Turing's formulation is the fundamental one. A more general compendium is Jean Van Heijenoort, *From Frege to Gödel: A Source Book in Mathematical Logic*, 1879–1931 (Cambridge, Massachusetts: Harvard University Press, 1967). Hofstadter's *Gödel, Escher, Bach* provides a more negotiable and enjoyable route for the general reader.

The paper in which Turing introduces his now famous test is 'Computing Machinery and Intelligence', *Mind*, Vol. LIX, No. 236 (1950). This article has been reproduced in several collections, for example, Edward Feigenbaum and Julian Feldman, eds, *Computers and Thought* (New York: McGraw-Hill, 1960). It is most recently, and widely, available in Douglas Hofstadter and Daniel Dennett, eds, *The Mind's I* (New York: Basic Books, 1981), which has been translated into several European languages and Japanese. Hofstadter and Dennett, however, removed two substantial portions of Turing's paper. The first, indicated by ellipses, is a basic description of the electronic digital computer and of the more general notion of discrete state machine. For the intended readers this discussion would be either unnecessary or too demandingly technical. The second, whose absence is not indicated in any way, is the final section of Turing's paper, titled 'Learning Machines'. While this section is perhaps mostly irrelevant to the concerns of Hofstadter's and Dennett's volume (and probably seemed more so in 1980), this section contains the 'skins of an onion' analogy, the suggestion that an intelligent computer should contain a random guess element, and most importantly, Turing's discussion of the 'child machine' learning approach, which has become such a fashionable alternative lately under the label 'connectionism', and his comparison of mutational/evolutionary processes to the computer science experimenter who tries out various initial states for the child machine. I discuss these topics in chs 9 and 10. *The Mind's I* also includes Hofstadter's brilliant 'Ant Fugue', Dennett's 'Where Am I?', and John Searle's 'Minds, Brains, and Programs', with which I shall contend in ch. 9. Modesty forbids that I mention that *The Mind's I* also contains the opening chapters of my own first novel, *Beyond Rejection* (New York: Random House/Bantam, 1980).

A related, and largely unknown, paper of Turing's, 'Intelligent Machinery', composed in 1948, appeared some years after his death in *Machine Intelligence 5*, eds Bernard Meltzer and Donald Michie (Edinburgh: Edinburgh University Press, 1969). Turing's abstract states:

The possible ways in which machinery might be made to show intelligent behaviour are discussed. The analogy with the human brain is used as a guiding principle. It is pointed out that the potentialities of the human intelligence can only be realized if suitable education is provided. The idea of an unorganized machine is defined, and it is suggested that the infant human cortex is of this nature.

This paper captures well the spirit and rhetoric of current day connectionist talk of 'neural nets'; there is also discussion of 'b-type unorganized machines' that reminds one of current talk of 'massively parallel distributed processing'. I make considerable use of this paper in chs 9 and 10.

2 Richard Dawkins, in *The Selfish Gene* (Oxford: Oxford University Press, 1976), coined 'meme' to mean the cultural equivalent of his selfish genes. As a blend of *mean* and *gene*, 'Dawkins' invention adds a Turingesque cast to older terms such as *idea* or *thought*. Dawkins called genes 'selfish' because it is they who play the ultimate competitive game of evolution: genes are self-replicating DNA mechanisms, who use plant and animal organisms as their 'survival machines'. Individual organisms are not the ultimate competitors because they always die, are even designed to do so, and they carry on reproductively and die to preserve their genes. A *meme* is an intellectual mechanism, a strand of mental computation, that a particular human may happen to introduce but which may well spread itself from mind to mind, invading an entire population and its offspring long after its initial host may be forgotten. Meme-ic evolution would seem to differ in two respects from genetic evolution: 1) memes can spread (or disappear) exponentially more quickly than genes, some taking minutes to infect a population, though doubt-less the most powerful and enduring ones take a generation or several; 2) since

memes spread primarily through the language of cognitively advanced, self-conscious creatures like us, memes pernicious to the individual or its group would seem less possible than genetic ones. However, the way in which self-destructive notions can captivate individuals and nations gives us pause about the second point. The completely unexpected, virulent spread of computer viruses also suggests that we may have, unwittingly, acquired a number of immunities that computers are only beginning to acquire.

3 The Thomas Kuhn quote is from p. 145 of *The Structure of Scientific Revolutions*; familiarly, it expresses Kuhn's most noted thesis. Kuhn's examples are from astronomy, chemistry, and physics. He supposes that 'normal science' is constituted by a 'paradigm', a shared and narrow sense of what is known, of what puzzles still exist, and of what constitutes proper experimental design and procedure. Before normal science we have an unfocused, unprofessional theorizing and data collection under many banners, incoherently milling about in all too many directions (physics before Newton, chemistry before Lavoisier, etc.). Normal science is necessary to the rapid advance in precise knowledge within the overarching paradigm: the accumulated data, however, does *not* determine that the paradigm is the only possible one. But scientists (quite properly) will pay little attention to any other possibility unless anomalous, paradigm-incompatible data brew up a crisis *and* an alternative paradigm is vigorously touted that explains the anomalies and subsumes the older paradigm's achievements within a larger, reoriented framework. I cast Ludwig Wittgenstein as the most ingenious and implacable anomaly collector (if not confessor); Gödel and Turing, of course, put forward the central logico-mathematical anomaly. Turing is the creator and touter of the new paradigm. Since, as I suggest at various points, cognitive 'science' (as distinct from relevant parts of logic and mathematics) perhaps had no paradigm before Turing, *et al.*, I can also call Wittgenstein a 'cognitive naturalist'. The two sentences quoted from Wittgenstein are from his *Philosophical Investigations* (Oxford: Basil Blackwell, 1958), p. 209.

5 Dark Glass and Shattered Mirrors

1 This chapter begins with quotes from numbered paragraphs 90, 114, and 115 of *Philosophical Investigations*. Russell gives many accounts of his relationship with Wittgenstein. His letters to Lady Ottoline Morrell vividly record the effect on him of Wittgenstein's criticism; indeed Russell suppressed the book he labored on in 1913–14 and it only appeared long after Russell's death. Turing's wry comment 'if I were not *here*, I should say aleph-nought' appears in Cora Diamond, ed., *Wittgenstein's Lectures on the Foundations of Mathematics, Cambridge 1939* (Ithaca, New York: Cornell University Press, 1976), p. 31. The Diamond edited volume is subtitled 'from the notes of R. G. Bosanquet, Norman Malcolm, Rush Rhees, and Yorick Smythies'; also present at the lectures were D. A. T. Gasking, A. C. Jackson, J. N. Finlay, Casimer

Lewy, Marya Lutman-Kokoszynska, Stephen Toulmin, Alister Watson, John Wisdom, and G. H. von Wright.

2 Wittgenstein quotes St Augustine's account of child language learning, and introduces his own antithetical *five red apple* account of language as use, in the first numbered paragraph of *Investigations*. His probing of 'pointing to' this and that aspect is from *Investigations*, paragraph 33. Wittgenstein's stress on the *use*, rather than *meaning*, of words, on what words and setences *do*, as opposed to reference and truth conditions was the central slogan, repeated in two popular books and a number of learned papers, of the most well-known and influential British linguist of the 1930s, J. R. Firth (see his *The Tongues of Men and Speech*, Westport, Connecticut: Greenwood Press, 1987); Firth and Bronislaw Malinowski are credited as the founders of the London School of Linguistics.

3 Wittgenstein's stress on the inexplicability of child language acquisition has become a central theme of recent linguistics, evocatively labeled 'the poverty of the stimulus argument' by Noam Chomsky and others who suggest that the only *explanation* for the great speed and success with which the child, who often receives virtually no teaching or correction, comes to speak and hear a natural language is that the child has an innate language acquisition device. In ch. 10 I go into more detail about this. Both Turing (in a general sense) and Chomsky, *et al.*, want to tease out *what* this device presupposes about human languages; from the point of view of Wittgenstein's *Investigations* such *explanations* are something beyond view, for Wittgenstein wants, as he often remarks, to describe the contradictions of ordinary experience *before* any such explanatory attempts are made. Similarly, Wittgenstein suggests that Augustine, and others, unwarrantably suppose that one can 'point to the colour, shape, number, etc.'. Children do, in fact, generalize in quite surprisingly specific ways, ways which fit the age-old characteristics of the physical objects and the linguistic possibilities in the sound stream that humans are heir to. Richard L. Gregory, for example, demonstrated that neonates a *few hours* after birth will flinch when a flat surface moves slowly toward their face: the infant innately connects visual changes with tactile expectations (see his *The Intelligent Eye*, London: Weidenfeld and Nicolson, 1970). Neonates also smile and coo in response to a face-sized circle with dots for eyes and a crescent moon smile *more quickly than* to their mother's face.

4 Wittgenstein's claim that there are no sufficient and necessary characteristics for something being a 'game' but rather a 'family of resemblances' in which certain examples are stereotypical is often referenced in what has become a substantial area of research. 'Furniture' is a *superordinate* category in that people remember it in terms of stereotyped '*ordinate*' instances such as 'table', 'chair', 'sofa'; 'settee', 'ottoman', and 'loveseat' are *subordinate* to sofa. See E. Rosch, 'Principles of Categorization', in E. Rosch and B. Lloyd, eds, *Cognition and Categorization* (Hillsdale, New Jersey: Lawrence Erlbaum, 1978).

6 The Contradictions of the Public World

1 The quotes that preface this chapter are from *Investigations* paragraphs 618 and 619 and p. 228. Freud's engaging comment that philosophy (i.e. solipsistic doubt) is caricatured by paranoia (and religion by compulsion neurosis) can be found in his *Collected Papers* (London: Hogarth Press, 1950), p. 94.

2 Russell's lectures on logical atomism can be found in Robert Marsh, ed., *Essays on Logic and Knowledge* (London: Unwin Hyman, 1977). G. E. Moore's 'Proof of an External World' is in his *Philosophical Papers* (London: Allen and Unwin, 1959).

3 Wittgenstein's introduction of the private language diary are *Investigations* paragraphs 258–9; the suggestion that sensation S might be correlated with a rise in blood pressure is in paragraph 270. The comment that he obeys the rule 'blindly' is paragraphs 219–21. On p. 231 we find Wittgenstein's crucial claim that his investigations have no interest in what *natural science* might discover by way of explaining the many peculiarities and contradictions of everyday experience (you may compare this with the corresponding quote from Chomsky toward the end of ch. 10 in this volume in which Chomsky refers to his own linguistic investigations as natural science). The duck/rabbit quote is from *Investigations*, pp. 198–200; the comment about seeing other humans as automata is paragraph 420; Wittgenstein's specification of the criteria for the truth of a confession appears on p. 222.

4 You should be aware that my reading of Wittgenstein's *Investigations* is somewhat unusual in its emphasis, particularly with respect to the so-called 'private language argument' and rule-following. Professor Saul Kripke, in *On Wittgenstein on Following a Rule* (Cambridge, Massachusetts: Harvard University Press, 1984), has proposed that Wittgenstein's discussion of keeping a diary for recording incidences of the 'private sensation' S is a part of his discussion of the person who goes on, after counting 2, 4 . . . 98, 100, to 104, 108, and so on. I find it impossible not to connect this rule following discussion with that between Wittgenstein and Turing that I quote in ch. 7 in which Wittgenstein demands Turing distinguish between what is 'natural' and what is 'mathematically determined'. Wittgenstein's point, I take it, is that it is *human nature* to go on 102, 104, 106 . . . (and it is fortunate that this is so); however, the way the rule is *conventionally* explained does not exclude the bizzare behavior that Wittgenstein mentions. Wittgenstein's skepticism about rule-following is skepticism about the claim that mere convention, mere learning or instruction, can account for the manifest fact that humans *do not go wrong* in the ways that Wittgenstein imagines they might.

5 The Yeats poem is 'The Circus Animals' Desertion'; Yeats of course refers to himself but as poet captures a human plight and certainly Wittgenstein's.

7 Turing and Wittgenstein

1 The quote that begins the chapter is from lecture XXII, pp. 212–13, *Wittgenstein's Lectures on the Foundations of Mathematics*; the second quote is from lecture XIX, pp. 186–7. See also Marvin Minsky, *The Society of Mind* (New York: Simon and Schuster, 1988). Steven Levy's *Hackers* (New York: Dell, 1985) provides an engaging account of the Massachusetts Institute of Technology students (originally, members of MIT's Model Railway Train Society) who were first to so fall under the spell of computers that they effectively formed a new ethic and social system for living with them. This first generation believed all programs were free and common property (they somehow obtained keys to every room at MIT and access to every machine). There were no women hackers, nor did they even have female friends, and their favorite reading was E. E. Smith's *Lensmen* novel series, the most unrelievedly and naively elitist and male chauvinist saga in the history of science fiction. When this first generation of hackers moved to California they acquired female companions, software corporations, and a propertied interest in privacy and software ownership.

While it seems clear that Wittgenstein and Turing saw little of each other apart from the lecture series, one feels a voyeuristic *frisson* hearing the following passage from Lecture XVII on one-to-one correlations from *Wittgenstein's Lectures on the Philosophy of Mathematics*:

It is the relationship existing between two things if the one is *a* and the other is *b*, the relation which they stand in by one being Turing and the other being Wittgenstein. It is surprising that this should be called a relation; one is inclined to say, 'And now let's have a relation.'

What is the relation between two things if one eats Turing and the other eats Wittgenstein? Say two lions. Or two dogs, one of whom bit Turing and the other of whom bit Wittgenstein. There might be several couples of dogs for which this was true. So that T and W – or *a* and *b* – would be, as it were, the two 'test bodies' for *x* and *y*, and *x* and *y* would have their relation by the one biting *a* and the other biting *b*. We could go about and ask, 'Which two dogs have this relation?'

But what if I said, 'What two people have the relation of the one eating himself and the other eating himself?' – if all people eat themselves? What would be the test here? You couldn't use this – the one eats himself and the other eats himself – to find out if two people had the relation.

You could use this – the one bites Wittgenstein and the other Turing – to correlate two classes. How would you use this relation – the one loves himself and the other loves himself – to find out whether they had the same number?

8 Information Storms

1 The chapter initial quote is from 'Computing Machinery and Intelligence', Section 4, Feigenbaum and Feldman, p. 16.

2 The Chomsky quote is from the first paragraph of his *Syntactic Structures* (The Hague: Mouton, 1957); my reason for introducing it becomes clearer in ch. 10.

3 Malcolm MacPhail's account of aiding Turing in building relay switches at Princeton is in Hodges, p. 138.

4 The account of Turing's career and the Enigma Machine are from Andrew Hodges' excellent biography, *Alan Turing: The Enigma* (New York: Simon and Schuster, 1983). Through this chapter and the next I quote from Turing's 'Computing Machinery and Intelligence' as it appears in Feigenbaum and Feldman and from 'Intelligent Machinery' as it appears in Meltzer and Michie (see the notes for ch. 4 for the full citation). Since 'Computing Machinery and Intelligence' is available in many other places, I shall indicate the number from the paper as well as the page number in Feigenbaum and Feldman. For reasons of availability I have preferred quoting from this paper rather than 'Intelligent Machinery' when the same point is made. The reader is reminded of the discussion in the notes for ch. 4, particularly the penultimate paragraph: 'Intelligent Machinery' provides much of the basis for maintaining that Turing had thought much about connectionism ('unorganized machines') and massively parallel distributed processing (collections of 'type-b organized machines').

9 The Imitation Game

1 The chapter headers are from Wittgenstein's *Investigations*, numbered paragraphs 259 and 260, and from Turing's 'Computing Machinery and Intelligence', Feigenbaum and Feldman, Section 6, p. 19, and Section 7, p. 35.

2 We have from Turing, 'Computing Machinery and Intelligence', Feigenbaum and Feldman, p. 11, Section 1; p. 24, Section 6, Objection 5; p. 27, Section 6, Objection 6; p. 30, Section 7; p. 22, Section 6, Objection 4; p. 31, Section 7; p. 34, Section 7; p. 31, Section 7.

3 The Hofstadter quote is from his *Metamathematical Themas* (New York: Basic Books, 1985), p. 443.

4 Ned Block's 'Troubles with Functionalism' is in his, ed., *Reading in the Philosophy of Psychology* (Cambridge, Massachusetts: Harvard University Press, 1985). John Searle's 'Minds, Brains, and Programs', *Behavioral and Brain Sciences* (1980) appears in book form as *Minds, Brains, and Science* (Cambridge, Massachusetts: Harvard University Press, 1985). Dennis Holding, *The Psychology of Chess Skill* (Hillsdale, New Jersey: Lawrence Erlbaum, 1985), provides the best available account of human and computer chess playing. Ian McEwan's *The Imitation Game* appeared as BBC's *Play For Today*, 24 April, 1980; it is available under that name along with some other

plays (Boston: Houghton Mifflin, 1982). The reader might enjoy Justin Leiber, *Can Animals and Machines Be Persons?* (Indianapolis, Indiana: Hackett, 1985); the notes provide some biographical information about Mary Godwin and the eighteenth-century male attitudes with which she contended. The quote about Turing's voice is from Hodges, p. 209.

The Linguistic Turn: The Child Program

1 The header quotes are Turing, 'Machine Intelligence', Meltzer and Michie, p. 14, and 'Computing Machinery and Intelligence', Feigenbaum and Feldman, p. 32, Section 7 (the mid-chapter quote is from the same section, p. 33).
2 Noam Chomsky, *Language and Knowledge: The Managua Lectures* (Cambridge, Massachusetts: MIT Press, 1988) gives a good overview of recent work in linguistics.
3 Susan Curtiss, *Genie: A Linguistic Study of a Modern-Day "Wild Child"* (New York: Academic Press, 1977) gives the relevant account.
4 See H. Terrace, *Nim: A Chimpanzee Who Learned Sign Language* (New York: Knopf, 1986). Ursula Bellugi and Edward Klima's *The Signs of Language* (Cambridge, Massachusetts: Harvard University Press, 1978) provides a fine account of human sign language, American Sign Language in particular.
5 Turing's 'penultimate word' is from 'Intelligent Machines', Meltzer and Michie, p. 17.
6 My source for Prigogine and his comments on Turing is Hodges, p. 564 (Hodges' account of Prigogine's conversations with Turing derive from Professor W. Byers Brown diary entries).
7 Dr Claudius appears in C. S. Forester, *Hornblower During the Crisis* (Boston: Little, Brown, 1967).

11 Stories of Consciousness

1 The chapter begins with quotes from La Mettrie, *Man a Machine*, p. 89; Douglas Hofstadter, *Metamathematical Themas*, p. 487; and Michael Gazzaniga and Joseph LeDoux, *The Integrated Mind* (New York: Plenum Press), pp. 150–1. In *The Origins of Consciousness in the Breakdown of the Bicameral Mind* (Boston: Houghton Mifflin, 1976), Julian Jaynes proposes that consciousness (in our sense of an inner life) came into existence for the ancient Greeks somewhere between the composition of the *Iliad* and the *Odyssey* – a product helped along by story-telling and the development of perspectival graphic representation. While Jaynes may be much too bold in his dating, and in other aspects of his theory, the notion of consciousness as an ongoing self-narration, one which, as Gazzaniga, *et al.*, show us, builds up a more reasonable account through self-observation than by running the whole show (recall Wittgenstein's remarks about the illusiveness of willing one's own actions). A 'Turing test-passing' computer might surely have related structures and problems. Such a

computer, perhaps operating with motoric equipment in the physical world, would need a kind of compact, ongoing 'self-story', both to communicate appropriately in human interchanges and to have a sense (in those terms) of what it was about more generally or when alone.

2 G. K. Chesterton, *The Victorian Age in Literature* (New York: Henry Holt, 1913), p. 134.

Index